Bootstrap Geologist

UNIVERSITY PRESS OF FLORIDA

Florida A&M University, Tallahassee
Florida Atlantic University, Boca Raton
Florida Gulf Coast University, Ft. Myers
Florida International University, Miami
Florida State University, Tallahassee
New College of Florida, Sarasota
University of Central Florida, Orlando
University of Florida, Gainesville
University of North Florida, Jacksonville
University of South Florida, Tampa
University of West Florida, Pensacola

Gene poses aboard USGS drill barge *T. W. Vaughan* while installing monitoring wells in south Biscayne Bay, 1995.

Bootstrap Geologist

My Life in Science

Gene Shinn

University Press of Florida

Gainesville | Tallahassee | Tampa | Boca Raton | Pensacola

Orlando | Miami | Jacksonville | Ft. Myers | Sarasota

This book may be available in an electronic edition.

18 17 16 15 14 13   6 5 4 3 2 1

Library of Congress Cataloging-in-Publication Data
Shinn, Eugene A.
Bootstrap geologist : my life in science / Gene Shinn.
p. cm.
Includes bibliographical references and index.
ISBN 978-0-8130-4436-1 (alk. paper)
1. Shinn, Eugene A. 2. Geologists—United States—Biography. 3. Geology—
United States—History. I. Title.
QE22.S556A3 2013
551.092—dc23 [B]   2012039464

The University Press of Florida is the scholarly publishing agency for the State University
System of Florida, comprising Florida A&M University, Florida Atlantic University, Florida
Gulf Coast University, Florida International University, Florida State University, New
College of Florida, University of Central Florida, University of Florida, University
of North Florida, University of South Florida, and University of West Florida.

University Press of Florida
15 Northwest 15th Street
Gainesville, FL 32611-2079
http://www.upf.com

I dedicate this book to my wife, Patricia Shinn, who has been part
of my life and adventures since the beginning, and to my parents,
Ellsworth and Olivia Shinn (aka Papa-San and Mama-San),
who encouraged me to be adventurous and to work overseas
at the first opportunity.

Contents

Preface: A Modern Horror Story

As he pushed his scraggly flock down the dry wadi, the turbaned old goat herder puzzled over the strange metal tower behind the bordering dunes. Camel drivers had told him such towers were for mobile phones—whatever mobile phones were.

But the camel men knew, from their travels, all about cell phones and Internet cafés; the camel men roamed the Western Sahara from here in Mauritania as far north as Morocco. The wizened, barefooted Arab just shook his gray head. He had all he could do, just looking after his goats.

He rapped the backs of a few deserters with his spindly goat stick, keeping the errant animals in line, away from the ominous tower. Why did the towers need soldiers to guard them, way out here in the desert? He had learned to keep his distance, to avoid the armed soldiers who appeared whenever he and his goats came too close. Several of his flock had disappeared previously, when their natural curiosity led them to inspect the tower. A few other goats that he was able to reclaim behaved strangely thereafter, sickened, and died. The camel drivers relayed similar accounts—camels, goats, and birds, even people behaved strangely.

Was it Allah's punishment? Were the towers works of the Devil? American agents? Why were Arab soldiers guarding them? Not even a country man like the old goat herder noticed the fine spray that wafted westward, toward the Atlantic coast. It was invisible during the periodic dust storms, the harmattans, when the towers disappeared in the dust-choked winds that blasted westward.

The old goat herder could not have known that two thousand miles away, people in the Caribbean Islands were falling ill from a variety of diseases, that asthma was reaching epidemic proportions in the Lesser Antilles, or that the government of Cuba had accused the Americans of practicing bioterrorism. Foot-and-mouth disease was already afflicting the scrawny Caribbean cattle. Soon the robust-blooded herds of central Florida would be infected. There were outbreaks of new or rare diseases in the United States that were transmitted to humans as well as to livestock: West Nile virus, anthrax. People were dying!

And not just humans and their livestock. Coral reefs throughout the Caribbean had been dying for several years. Amphibians in Central American rain forests were succumbing to fast-spreading fungal diseases. Sea turtles expired mysteriously. Louisiana rice rotted from red stripe, a disease new to the United States. Citrus canker gnarled leaves and marred the surface of fruit in Florida grapefruit, orange, and lemon groves.

The patterns of the outbreaks puzzled American epidemiologists. The Centers for Disease Control would not confess that they were stumped, instead sticking to their first explanation: "The problems stemmed from overseas meat and produce smuggled into the country in the luggage of international air travelers."

•

Science fiction? Opening passages of the latest terrorism thriller? Yes. But could such a chilling scenario actually happen?

I wrote that scary introduction to a story about our research on African dust for *Reader's Digest*. It was only a few weeks before the horror of 9/11, and the anthrax letters followed soon thereafter. A general from Fort Detrick, Maryland, had already told me about anthrax. "It's a really tough bug," he said. "If someone put it on the Sahara Desert, it would hit us in about six days." He told me this after hearing my presentation on African dust. His words confirmed what I had written.

A geologist I met at a meeting in Scotland told me, "If you want attention that may help you with funding, forget the scientific literature. Publish first in *Reader's Digest*. That's what the decision makers read!" Something I would never have dreamed of, but I was desperate to keep our project moving. A terror-filled introduction just might do the trick. The truth is I *was* envisioning a terrorist attack. I really was concerned. Why? I had just read *Biohazard*, the 1999 book by Ken Alibek, the Russian defector and germ-warfare expert. Alibek confirmed that thousands of tons of weaponized anthrax, *Bacillus anthraces*, had been produced under his supervision. That was during the Cold War. Officially it was destroyed, but do we really know? Could some have found its way to western Africa? Would cellphone towers be the perfect cover for dispersal?

By then, my team of microbiologists had identified over one hundred species of live germs and fungal spores attached to African dust that had crossed the Atlantic. They had found the bug that causes red stripe disease in rice. Among the many species, they found lots of *Bacillus*, but thankfully no *Bacillus anthraces*. Their finding proved that the really nasty bug could survive the long trip. All that was needed was a way to release it during a raging dust storm. We knew this could be done, even before the anthrax letter attack that killed five U.S. citizens and sickened others. After those letters, I was convinced there was reason to worry.

I couldn't sleep, and it was especially frustrating. No one in authority seemed to want to deal with the real possibility. I also knew that I risked the scorn of high-browed scientists if I resorted to the *Reader's Digest*. I suppose it's good the magazine didn't publish the story. The editor liked it but said, "It's too long." *Digest* stories are short, but I lacked the skill to distill this convoluted story for a quick read.

•

I will tell the whole story later in this book. But first, before we attack this huge problem of African dust, let's get back to the beginning. This is really the life story of a successful self-taught geological scientist, initially handicapped by lack of advanced degrees, who rose to the top of his profession and received its highest award—the Twenhofel Medal. In relating the life of a "bootstrap geologist," the book chronicles life's many forks that can now be evaluated and appreciated in retrospect.

The book's main purpose is to tell some wonderful stories of earth science (much of it underwater) and to chronicle fifty years of working with the great, the grand, and the merely mad people in marine science. Writing about the life of a scientist, and the inner workings of science, in a reader-friendly way is challenging. For that reason, I decided to tell my story as simply as possible, taking care to avoid technical jargon where suitable. When not possible, jargon is explained in endnotes.

For the most part, I focus on the many adventures, and colleagues, associated with doing field research in remote and exotic places, and I include the conflicts with other scientists—often, armchair scientists—over interpretation and meaning of discoveries. I have always been attracted to difficult, important, large-scale problems, and the reader will see that I've relied more on what can be seen with the eye than on that which is calculated and modeled. Through my memories, I hope to show how science works.

I

Beginnings and Early Years

Birth of a Conch

EK, that's my father, dropped ether on the cotton pad over Mom's nose while Dr. DePoo brought me into what would become the Conch Republic. Being born in Key West automatically makes me a "saltwater conch" as opposed to those who move there. They are called "freshwater conchs." This conch's birth happened November 7, 1933, on Von Phister Street in Key West, Florida. The Great Depression was at its deepest, and the western Dust Bowl days were at full throttle. Only a week earlier, hurricane winds had sloshed a foot of Atlantic waters through our home.

My mother, Olivia Allen, was born in Columbus, Georgia, the eldest of eight. It sounds trite, but as in the song, she met my father in the five-and-dime store. He was a private stationed at Fort Benning, Georgia. Dad, formally named Ellsworth Kellogg Shinn, disliked his first name and for that reason became known to his friends as EK. Seventy years later he would be known in the family as Papa-San. EK was born on Valentine's Day in Rolla, Missouri, in 1910, but at age fourteen moved to Slidell, Louisiana, to be raised by his schoolteacher aunt. He graduated from Slidell High School in 1927 and joined the Army to further his study of radio and communications. Radio and any form of electronics were his first loves. Shooting squirrels for food and trapping muskrats for pelts along the Pearl River were close seconds. After the Army and its radio school, his first job was for the Lighthouse Service in Key West. That was 1930. He was radio operator on the lighthouse tender *Ivy* under Captain Cosgrove. His job—the crew called him Sparks—included installing and repairing two-way radios in lighthouses on both Florida coasts.

Key West was known for its high rate of tuberculosis, and because EK was so thin—119 pounds and six feet tall—the service required that he have a chest X-ray. Only one person in Key West had an X-ray machine. He was an interesting German doctor named Karl Tanzler Von Cosel. Just how interesting Von Cosel was would become apparent several years later.

Depression times being what they were, the Lighthouse Service soon tightened its belt and reduced services and personnel. There was a general reduction in force in 1933, known in government service as a "RIF." Fortunately, there was another government job involving radio available, but it was in Phoenix, Arizona. The job was a radio-communications position with the Civil Aeronautics Administration (CAA), later to become the Federal Aviation Administration.

The only problem was getting there. An automobile had not been needed in Key West. EK and Olivia went everywhere on bicycles. Life was easy and shoes were optional. Most Key Wester conchs lived on "grits and grunts." Grunts are ubiquitous little reef fish that could be caught under any dock or seawall with a simple hand line. Lobsters, known to conchs as crawfish, were abundant. Grits, as every southerner knows, is ground corn. Sometimes people jokingly called it Georgia ice cream.

To reach Phoenix, Dad bought an English-made Ford sedan from Ernest Hemingway. The emerging writer lived just down the street. The car was different—not the standard Ford black, but tan. It had a plaque on the dashboard that said, "Specially made for Ernest Hemingway." Ernest, as many readers know, was a notorious, self-absorbed, hard-drinking fellow.

During the early days, Key West remained more closely allied to Cuba than to the United States. Hemingway was, and remains, a hero in Cuba. Every schoolchild in Cuba knows of Hemingway and *The Old Man and the Sea*. Cuba's largest marina, still open for American yachts, is called the Hemingway Marina. Financier Henry Morrison Flagler's railroad had reached Key West in 1912. It connected, via ferryboats, to the rail system in Cuba. The original reason for the railroad was to bring coal to Key West so that cargo and military ships could refuel while cruising to and from the Gulf of Mexico or beyond. By the time Flagler's railroad reached Key West, most ships had converted to fuel oil, so instead of coal the railroad transported fruit and tourists from Florida and Cuba. Remains of the loading dock still exist at the north end of Duval Street where freight cars once boarded ferryboats headed to and from Cuba. The ferryboats were a seagoing connection to the rail system in Cuba.

Going north by automobile to reach mainland Florida required a long ferryboat trip. The ferryboat left from No Name Key and landed at Long Key. The remainder of the trip was via a narrow road with many rickety wooden bridges. Driving on the planked bridge roadways made a terrific racket. Flagler's railroad was swept away in the great Labor Day Hurricane of 1935. It happened just two years after we left, and EK would be eyewitness to the storm's devastation.

The trip from Key West to Phoenix is family legend with elements akin to Steinbeck's *The Grapes of Wrath*. On the relatively new but narrow Tamiami Trail, U.S. 41, and in the Everglades just west of Miami, we were sideswiped by another car. It destroyed our luggage trailer along with our few possessions. I was two weeks old and sleeping on the backseat. Seat belts and child seats did not exist then, but I survived. We limped into Columbus, Georgia, my mother's home, without the trailer. Grandfather Allen helped EK make the needed repairs. This English-made Ford was difficult to repair; I was told later that it had things like bolts with left-hand threads. Soon we were off again, but to survive and eat during the long trip to Phoenix, EK reluctantly pawned his beloved banjo. Later he pawned his typewriter. It was also very cold driving out west. Somewhere in Arizona, frostbite numbed Dad's big toe. It would bother him for years. The Shinn family arrived in Phoenix with fourteen cents! Fortunately, the new job was waiting and credit was available at local markets. Tarantulas were abundant but not considered edible. Needless to say, I remember nothing of this adventure, just the tall tales told and retold around the dinner table. I do know that up to then my birth certificate said "Baby Shinn." I was finally named Eugene in Phoenix. I am told that until I was properly named I was called "Squirt." Dad said that every time he picked me up, I squirted!

The Depression deepened, and our cash-poor government tightened its belt again. This time EK and family moved to a new job with a fledgling airline in New Jersey. A fellow named Eddie Rickenbacker was starting a new service called Eastern Airlines. EK did not particularly like Rickenbacker, and to make things worse, he hated New Jersey and Yankee attitudes. Having spent his teenage years paddling a pirogue in the swamps of Louisiana around Slidell, he felt he knew something about paddling. He was outraged when a New Jersey park ranger pulled him over for not rowing in the "proper manner." He was not sitting in the middle and using two oars. He paddled from the rear of the boat the way any respectable Cajun would paddle a pirogue deep in Louisiana's swamps.

I think Dad had problems with authority, which may explain why he did not get along with Eddie Rickenbacker. His dealings with Captain Eddie would lead to an interesting relationship many years later, long after Rickenbacker became a World War II hero. I must have inherited, or maybe absorbed, EK's suspicion of authority. Suspicion of authority did not mean breaking laws or harming others. He lived by the Golden Rule. It was more like, "Don't tell me how to run my life or paddle my boat."

Soon we were in Columbus, Georgia, again. EK worked in a radio-repair shop, and when not fixing radios he was a radio announcer for station WRBL. He played records over the air before the term "disk jockey" was invented. Until his passing in 1996 at age eighty-seven, EK remained what is called a "Ham Operator," the universal term for amateur radio operators. He had been on the radio since he was fourteen years old and was a Pioneer radio ham with the call letters W4BIC.

Morse code was second nature to him, and he still preferred code even after voice came in. He could send sixty words a minute! He even had a set in his car and used a "Bug Key" that sat next to him on the front seat. Bug Keys work side-to-side between thumb and forefinger and are faster than the conventional key that only goes up and down. He once told me how exciting it was when he first heard a human voice come through his headphones. "It scared me!" he said.

EK played the guitar and banjo and in due course replaced those he had pawned. Eventually, the economy improved a little and in early 1935 he was reinstated with the CAA, but this time the job was in Titusville, Florida. I don't remember that move either, but I do know that he had another sideline to make extra money—another talent he learned in the swamps near Slidell. He ran a trapline and sold rabbits, 'coons, and possums. They were sold mainly in the poor black neighborhood in Titusville, where he acquired the nickname "de Possum man." His stories of removing stinky live skunks from steel traps without being sprayed were family favorites.

The great Labor Day Hurricane struck the Florida Keys with a vengeance while we lived in Titusville. The deadly storm killed more than four hundred civilians and World War I veterans who were sent to build roads and bridges as part of a Federal Economic Recovery Act project. It was a Category 5 hurricane, and the barometric pressure recorded at Craig Key remains the lowest ever recorded in the United States.

The evacuation train sent to save the veterans and citizens of the Keys was late because of bureaucratic screw-ups and miscommunications. It reached Islamorada just before a gigantic wave overturned all the passenger and freight cars. Only the heavy locomotive and its attached car remained on the track as storm waves washed completely across the island. Some blamed President Roosevelt and his New Deal for the bureaucratic mess. In fact, it was caused by a lack of communications within the railroad and by human bungling.

After the storm, EK, along with others, was sent to Islamorada to set up ham radio for communications. There were no phones, and ham radio operators were always in demand during and after disasters. Even today, they still play important roles in times of national disaster, especially in third-world countries.

As one might expect, EK was witness to the terrible death and destruction in the middle Florida Keys. He saw all the gruesome sights that Hemingway and others so vividly describe in their writings.[1]

Return to Key West and Weirdness

After Titusville, the family moved to Atlanta, and then to Columbus, where I attended kindergarten. Next it was Augusta, where we lived in a farmhouse on the outskirts of town. The house was near a swamp and also near the airfield where EK worked as a radio communicator and meteorologist. We did not have an indoor toilet, and my memories of the place center around trips to the "outhouse" where corncobs and old Sears catalogues served their purpose. And there were spiders! The farmhouse was on a hill

and a few miles from a creek that ran through a swamp. We did a lot of
bream fishing in the creek, and Dad shot squirrels with his .22-caliber rifle.
We ate a lot of squirrels and fish. I remember gagging on small fish bones,
but a swallow of bread usually washed them down. Dad once extracted one
from Mom's throat with long-nosed pliers.

Part of EK's job was communicating weather conditions to pilots, and in
the process he became an accomplished and respected "bootstrap" weath-
erman. That skill would play a larger role in the future.

Soon we were headed south again, back to Key West, where I attended
the first grade. A lot of that period I do remember. I am told I spoke Spanish because my playmates were mainly Spanish. I do not remember what
Spanish I learned—I think they were mainly curse words that have long
been forgotten.

During this time in Key West, EK worked the midwatch, often called
the graveyard shift, at the CAA radio station. Because of his job, he needed
to sleep during the day. Sleep was difficult because the Cuban lady in the
house next door—houses were about six feet apart—turned her radio up
full blast. Using his electronic skills, EK constructed a radio-frequency oscillator. When the blaring radio went on, he would tune the oscillator to the
same frequency, causing the radio to make horrific sounds. After a tirade
of Spanish curses, the lady would find a different station and EK would
quickly tune in the new station. Soon the radio would be turned off in
disgust! The radio repair shop could never find anything wrong with her
set—but Dad slept better. Being the devilish creative person he was, I'm
sure he relished the fun he was having.

One memory is vivid—a loud boom that shook the island! It was around
midnight, and the sound woke most of the city. Dr. Von Cosel, the same
"interesting" doctor who took my father's chest X-ray, had dynamited the
crypt of his departed wife to make it appear he had just stolen her body. The
truth was that he had taken the body nine years earlier on the night she was
buried.

The only person who knew he had the body, aside from the doctor
who delivered me, was the mother of his wife. She lived across the street
from us, and of course my father knew her! Von Cosel was sixty when
he married twenty-two-year-old Elena. Von Cosel was receiving a World
War I pension from Germany when they married, so he was doing well.

Unfortunately, Elena died of "consumption" soon after they were married. Consumption was what they often called tuberculosis, or TB, in those days. Elena's mother knew Von Cosel had her daughter's body but told no one because Von Cosel was bribing her to keep quiet. However, sometime before the start of World War II, Von Cosel stopped receiving his pension and could no longer pay the bribes. Elena's mother threatened to tell the police, so Von Cosel dynamited the crypt in the middle of the night and fabricated the story that he had just taken her body. He claimed he had been working all those years on ways to resuscitate her. Apparently, he truly thought he could bring her back in a Frankenstein sort of way. To prove how serious, or crazy, or smart he was—Von Cosel is said to have held nine college degrees—he had even built an airplane to fly her away once she came back. I remember the airplane. The wings had not yet been attached. The nose with the big radial engine protruded from beneath a tin-roofed room of a long storage shed.

Von Cosel had also built his own pipe organ. He often played the organ while dressed in a white suit, much as in a scene from a Vincent Price horror film. To complete the bizarre picture, he had made a plaster death mask of Elena. It was mounted on the wall beside the organ so she could look down as he played and worked the base pedals with his feet. Being the strange place Key West was, and remains today, the authorities actually believed he might bring her back from the dead. They allowed him to keep the body! It could only happen in Key West—or maybe Transylvania.

Soon Von Cosel made a deal, and a Spanish film crew arrived from Havana to make a feature story of his experiments. They set up some elaborate backdrops, and tourists flocked to Key West to watch the filming and glimpse the body, for a fee of course. EK had no prior knowledge of the body, but after the story broke he did hear a lot from Elena's mother. He also heard from Von Cosel, who enlisted EK to take still photographs while the Spanish film crew worked. Dad's photographs included Von Cosel playing the organ and making sparks with his large Tesla coil. Most of Dad's photos were printed in his darkroom and made into picture postcards that Von Cosel sold to the tourists who mainly came from Miami. It was reported that on one weekend as many as six thousand people drove down to observe the body and the film crew at work.

Just like in the old Frankenstein movies, Von Cosel did experiments

Von Cosel with Tesla coil, which, in a Frankenstein-like way, was designed to provide electrical shocks to awaken his long-dead wife. Photo by E. K. Shinn.

with electricity. He had constructed a huge Tesla coil that was over six feet tall and had a mirror-like silver ball on top. Sparks shot from the silver ball. When he electrified the body, some tourists were said to have seen Elena's body twitch! Apparently those reports were enough to convince authorities that he just might succeed. That would really put Key West on the map, and the city needed the money! I don't remember how long the authorities went along with this ruse, but the film crew eventually returned to Cuba with a film "in the can." If ever found, that film would really be worth something today. I have read countless accounts of the Von Cosel affair but have never seen the making of the movie mentioned.[2]

The authorities eventually decided Elena was not coming back and confiscated the body. Von Cosel was arrested on October 8, 1940, and sent to Miami for trial. The *Miami Herald* purchased most of my father's unique photographs for a princely sum of $140—a lot of money in 1940! Every so often I come across yet another rendition of how it all happened. The stories are usually published around Halloween and use the photographs I know so well. The irony of it all was that Von Cosel won the case and was exonerated on the grounds that Elena was his property to do with as he chose. Imagine that today! After the trial, Von Cosel went back to Key West, gathered up his possessions, and using the fuselage of the airplane like a trailer, drove off to Zephyrhills, Florida. To my knowledge, he was never seen again. The shocker came later. Careful examination of the body revealed that Von Cosel had been sleeping with the highly reconstructed body during all those years. Love is strange.

Moving On

Our next move took us to Georgia. The war had started and EK was still with the CAA serving as radio operator and part-time meteorologist. While in Atlanta he taught basic meteorology to young CAA recruits, and his duties included sending up weather balloons to determine speed and direction of "winds aloft." Weather data were then transmitted to aircraft via radio. Because of the war, most CAA recruits were women. One recruit was my aunt Virginia, my mother's younger sister and the youngest of eight siblings. Their family relationship of course had to be kept secret.

Although EK had ranked as an "expert marksman" in the Army, his poor eyesight mandated eyeglasses. Eyeglasses and the wartime necessity of his job kept him out of the war in Europe or the Pacific. His knowledge of meteorology would play an even larger role later in life. He had never forgotten the aftermath of the Labor Day Storm in the Florida Keys, and many more hurricanes would affect the Shinns' lives.

Settling Down in Miami

The Shinn family eventually settled in Miami, where I attended school from grade 3 and on to the University of Miami. I graduated during midyear in 1957. I had also graduated from high school in midyear in 1953. Why midyear? I failed grade 6A and had to repeat that half semester. Report cards often said, "Does not follow instructions" or, "Does not come to class prepared." Although failure was blamed on our gypsy-like lifestyle, I think it was just inattention. Today, we would probably give it a medical term, like attention deficit disorder. Of course, being an only child, it was natural that I was spoiled rotten. Teachers always seemed to know I was an only child.

Settling in Miami did not stop the moving. We moved all over the growing city and suburbs. EK could not afford to purchase a home, so we rented. First it was a motel on Biscayne Bay in North Miami. I remember how the motel's owner kept me supplied with BBs to shoot invading land crabs with my Red Ryder BB gun. Crabs continually swarmed inland from the mangroves. They were everywhere, eating everything except sandspurs. Sandspurs seemed to be the most common native plant in Miami back then. Maybe it just seemed that way because I always went barefoot. I hated shoes, and most other boys my age also went barefoot.

EK eventually rented a home in North Miami. It too was near the bay and had abundant citrus trees in the backyard. While living there, I learned to gig mullet in nearby canals. A gig is a kind of spear on a wooden pole with three barbed prongs. I spent a lot of time patrolling bridges and canal banks searching for schools of mullet. From that part of town we moved to a suburb called El Portal. EK made a small smokehouse in the backyard in which he smoked fish that he sometimes sold to a local bar. "Salty smoked fish and beer go hand-in-hand," he said. I preferred water!

Next it was a trailer park on Little River Canal just off Seventy-Ninth Street. Dad built a screened-in canvas-topped wood deck on one side of the trailer. The deck hung a few feet over the water and had a trapdoor. The trapdoor saved long walks to the communal bathroom at night. I was still in grade school but had a rowboat in which I roamed the canal, and caught fish. That was when I learned to swim.

After the riverside trailer park we moved to a trailer park on Thirty-Sixth

Street a few miles east of the Miami Airport. I was growing up and was then attending Allapattah grade school. Our trailer shook as if we were having an earthquake when the large four-engine cargo and commercial planes took off. That was before jets, so they remained low and loud when the throbbing propellers thundered overhead. Our trailer was across the street from a large wooded area where I spent much of my time. By then I was into snakes. I kept dozens of water snakes in cages in the nearby woods, but sometimes they were under the trailer. Fortunately, about three blocks away there was a small company called Florida Reptile that made snake-leather belts and pocketbooks—such items were popular with northern tourists at the time. I sold them snakes for seventy-five cents a foot. I also learned to tan snakeskin and even made a few wallets and snakeskin-covered belts myself.

"Gene, come quick," a trailer park neighbor once summoned me, over to a trailer on the other side of the park. A man with poor eyesight had been cleaning a small patch of grass next to his trailer and reached down to remove a stick. The stick was a snake that immediately bit his hand. At that time, most people considered all snakes poisonous—a popular saying was that "the only good snake is a dead snake." Yes, he killed this one and then used a razor blade to cut the bites and suck out blood—standard practice for treating poisonous snakebites back then. He was very relieved when I told him the dead snake was not a water moccasin but a harmless banded water snake. "What is a water snake doing here so far from the water?" everyone asked. I kept a straight face, as it was clearly one of my wayward pets.

On another occasion, I was walking home from school when I encountered two small children dragging a small cage down the street. They were about a block from my trailer. I recognized the cage—it was mine—and it contained three small cottonmouth water moccasins! I quickly confiscated the cage. That episode was frightening. I never kept poisonous snakes after that.

It was during this period of trailer park living that young men of Hawaiian descent would regularly appear at local food stores. Big-chain grocery stores were rare. These fellows would stand in front of the stores and do yo-yo tricks, and all of us kids would gather around to watch. I got caught up in the Duncan yo-yo craze and learned every trick in the book, but the one thing I could never do was work two yo-yos—one in each hand—at the same time. The Hawaiian guys had us all beat. I eventually worked my way up to a black yo-yo on which the Hawaiian fellow carved a palm tree using his pocketknife. It was expensive and cost about fifty cents! This one also had some small fake diamonds inlaid on each side. Of course, it was made of wood. Plastic yo-yos were still way in the future.

One weekend, I took the city bus to downtown Miami—fare was ten cents—and entered my first and last yo-yo contest. We had to do tricks, but the telling, and most difficult, maneuver was to do loop-de-loops—not just once but continuously. I made it up to about 365 continuous loop-de-loops when the string became so tangled I had to stop. The problem is that each time the yo-yo makes a circuit around your finger, it twists the string one turn. The string gets tighter and tighter. A kid who did 500 loop-de-loops won the contest. I remember my father saying, "If you put all that energy into something useful, you could really make something of yourself." I still like to yo-yo and keep several in my office. Companies give them away as advertising gimmicks at national shows like the annual convention of the American Association of Petroleum Geologists. I use them to demonstrate sea-level fluctuations.

Everglades Flooding, Snakes, and the 1947 Hurricane

We were living in the trailer park east of Miami Airport when the hurricane of 1947 flooded much of Miami. West Miami and Hialeah went under one to three feet of water, and the Everglades were under several feet of water in some places. Snakes fleeing the water concentrated on any high ground, especially big diamondback rattlers. They got fat on the rabbits that also fled to high ground. Big ones made sufficient leather for large pocketbooks. I really got into snake catching when the high waters came. By then, I knew every snake in south Florida. The storm flooding changed south Florida, and the Corps of Engineers began the task of digging canals to drain the glades. EK said the reason for so much rain was because they had seeded the storm with silver-iodide crystals.

Our next move was to South Miami, where Mom and Dad managed a motel—the Patio Courts—on Southwest Eighth Street. The rent was free, and I had a motel unit all to myself. Of course, it was the noisiest unit,

Gene as a fourteen-year-old snake catcher with his homemade canoe in the Everglades.

closest to the road. EK still worked for the CAA while Mom worked very hard cleaning rooms and sheets. There were couples that checked in after dark on Saturday nights and left before daylight. Dad called them "Saturday-night specials."

Finally, we rented a home just a block to the north and just off Southwest Eighth Street. By then I had raised a burrowing owl I took from a nest and, of course, I still had a pet snake. It was a six-foot-long boa constrictor. One night, the snake got into the owl's cage. When daybreak came there were only a few feathers, and I found the snake in the yard with a huge bulge in its middle.

Next we rented a home in Hialeah, a very solid square box made of poured concrete and located not far from the famous Hialeah racetrack. They called the subdivision "The Sundecks." The houses had solid concrete roofs with a solid-cement five-foot-high wall around the roof perimeter. It served as a sundeck and was a great place to play. On one occasion, the eye of a hurricane—storms were not yet being named—passed directly over. The concrete home was like a fortress and didn't even shudder. The warm humidity and smells in the calm of the eye were unusual, and it also brought small exotic birds from Cuba.

EK finally purchased his first home. It too was in Hialeah, not far from The Sundecks. During this time, EK was promoted to branch chief of what by then had been renamed the Federal Aviation Administration (FAA). He did not inspect airplanes or work in the control tower; instead, he ran the communications center at Miami International Airport, where he managed about one hundred radio and Teletype operators. The communications division was separate from the air-traffic controllers, and in fact there was some animosity between the two groups. Air-traffic controllers got all the public attention and dealt with planes about to land. Communications dealt with and provided weather reports to commercial airlines while they were en route to their various destinations. Working as branch chief at Miami International, EK once again found himself dealing with the same Eddie Rickenbacker who had been his boss in New Jersey. Eastern Airlines was flourishing, and its headquarters were at Miami International. EK said one of Rickenbacker's stated goals was to privatize—I think the word was *eliminate*—the FAA. Eddie called the FAA "a monkey wrench in the wheels of progress." Although he tried, he didn't succeed.

EK continued his involvement with weather and became a friend of Grady Norton of the Florida Hurricane Service. Norton was the granddaddy of all Miami weathermen and the hurricane expert's expert! After the disastrous flooding by the particularly wet hurricane in 1947, EK (representing the FAA), Norton (representing the Weather Service), and some Army Corps of Engineers officers prepared the document that created the National Hurricane Center. The 1947 storm, as EK had told me, was the first to be seeded with silver-iodide crystals. Seeding promoted rain and thus reduced wind velocity. Maybe that caused the flooding? I don't know. I do know that the government "officially" stopped seeding storms thereafter, and the disastrous flooding prompted the Corps of Engineers to construct an elaborate drainage system to prevent future flooding. It worked, as there were no more serious flooding events in Miami. Ironically, forty-five years later, the Corps began the billion-dollar process of dismantling the system. Miami was using water faster than it could be delivered, and Everglades National Park was running dry during droughts!

Making Music with High School Sweetheart

When we were living in the trailer park on Thirty-Sixth Street, I had taken accordion lessons. That lasted about a year. Like most kids, I hated to practice, and besides, it was just the "squeezebox and me." After hitting the wall with the squeezebox, I joined the Miami Jackson High School beginners' band. I wanted to play clarinet. Unfortunately, the reed mouthpiece vibrated on my front teeth and gave me goose bumps. It was worse than rubbing my fingernails on a chalkboard. I gave up after one week. In desperation, I switched to drums and took private lessons from the band director. That was a smart move. I was hooked!

My musical career began at Miami Jackson. What started as a hobby and fun pastime led to meeting my wife and getting a college scholarship. Even though I could not play the clarinet, Patricia Moltz could, and we soon became high school sweethearts. I made senior band in one year, and music certainly kept me out of trouble. While my friends tuned hot rods, I marched with the band until sunset.

Our marching band at Miami Jackson was exceptional, and we won the national High School Marching Band championship in New York in 1948. We were a new kind of band for the South. Instead of a slow military cadence, we marched fast—almost ran—and danced while playing swing music. We were famous for the "St. Louis Blues March," and we always had a trumpet player who would step out of ranks and play a jazz riff. Our popularity led to yearly trips to Havana for the Sugar Festival Parade.

Once we performed the halftime show for the annual All-Star Game at Soldier Field in Chicago. Before the game was over, we were already boarding two chartered DC-3s and headed south for the Dominican Republic to play a private concert for President Trujillo. Having learned photography from EK, I became the high school newspaper photographer, contributed photos to the school yearbook, and was also president of our photography club. I took lots of pictures during our band travels, and when I stepped forward with my old Speed Graphic during Trujillo's concert, a burly guard snatched it from my hands. After deciding it was not a gun, he returned it and I shot the photo. Trujillo didn't flinch when the flashbulb went off. He was admiring—I should say ogling—the young nubile majorettes dancing a few feet away.

In our home there was the ever-present enlarger and photographic darkroom. The equipment was either in bathrooms or in small closets. Looking back on it all, it is a miracle I never succumbed to all those chemicals, especially acetic acid, "stop bath," and "fixer." They are potent, and none of my little darkroom closets was ventilated. There was no air-conditioning in any of our homes at that time, and there was none in Miami's public schools! Air-conditioning would not come to schools in south Florida until much later.

Drum Beating

Besides the banjo, guitar, and photography, EK had also taken up the drums. He was not very good, but he could keep the beat. He had a minimal drum set—bass drum, snare, and one cymbal. He played with some small dance-band groups for extra change.

Like my father, I played in pickup dance bands around Miami, especially on South Miami Beach long before its rebirth as Miami's primo Art Deco District. We played a lot of Jewish music, polkas, and of course rumbas. Ever hear the "Miami Beach Rhumba"? I got sick of it because I played it so many times. I felt the same way about "Moon over Miami." Then, South Beach was where less-affluent New Yorkers migrated for warmth during Brooklyn winters. The more-affluent "snowbirds," as we called them, migrated to North Miami Beach, meaning any place north of Lincoln Road.

New Yorkers were often the brunt of Jewish stand-up comedians. Punch lines depended on what part of the Beach they were playing. Stand-up comedians would come in during breaks and tell Jewish jokes. My job was to accent the punch lines with drum flourishes or cymbal crashes—just like in burlesque shows! If we were in South Beach, the jokes were mainly about New Yorkers in North Beach. When they performed in North Beach, the jokes were about those on South Beach. South Beach would play another role later in life, when I would drive there every day for fifteen years to catch a boat to my office.

One night while playing in a smoky bar, I had a visitor. I had landed this job with a pretty good small band, known to musicians as a combo, on Coral Way in South Miami. The job required long hours late at night, and like all late-night bars, it was smoke filled. During a break, two older burly men approached the band and bought us a drink. "You fellows Union members?" My antennae shot straight up! This was something we had been dreading. These men sounded just a little threatening. We admitted we were not members. Couldn't lie. They would just ask to see our membership cards. "You fellows are too good to not be members. Why don't you come down to the Union Hall and we'll sign you up? You will get a lot more jobs if you are members."

I was thoroughly intimidated, so a few days later I drove down to the Union office on Biscayne Boulevard, where I was ushered into a small room. A rather burly fellow came in to see me. I looked at his hands. He was no musician! Those hands would not fit any musical instrument on this planet. Brass knuckles seemed more likely. My head was screaming "MOBSTER." "It will only cost you seventy-five dollars to join," he explained. I forked over the money. That was a lot of money. "There will also be annual dues." After joining I never received a single union job, so I let the dues lapse. That was my first, last, and only experience with the Musician's Union. Maybe things have changed since then.

Upon graduating from high school in midyear 1953, I received a music scholarship to the University of Miami, where I played in the UM Symphony for four years. The best thing about the University of Miami scholarship was that I didn't have to major in music. I had no clue what to study but knew it would not be music. All those nights playing in smoke-filled bars prompted that decision.

I was grateful that music had provided me with a scholarship. The major requirement to keep the scholarship was to practice with the University of Miami Symphony three days a week and play two concerts a month. I dearly wanted to be in the UM marching band—"The Band of the Hour"— but they had an abundance of drummers. In retrospect, the symphony was a much-needed cultural experience that rounded off some of my sharp edges and made people believe I was somewhat cultured. Many famous visiting soloists and conductors traveled to sunny Miami to perform with the UM Symphony. I still remember playing a concert under the conductorship of Leopold Stokowski—Russian music with lots of cymbal crashing.

And then there was Brazilian-born Villa-Lobos. We performed some of his compositions that had difficult time changes, but best of all, I played the snare drum in Ravel's *Boléro*. That's dum-da-da-da-dum, da-da-da-dum for seventeen crescendoing minutes. Halfway through, he walked off the stage for a while and left me leading seventy-five musicians until he returned. It was a theatrical trick, of course, but I was nervous and will never forget it. The *Miami Herald* music critic even noted my performance the next day. At that time, 1953 to 1957, we were the major cultural highlight of that little "backwoods" town called Miami.[3]

While I was on scholarship, chemistry appealed to me, probably because of my Gilbert chemistry sets. I knew the smells and tastes of most of the chemicals they provided. I had made a lot of gunpowder! That was, after all, the main reason boys liked chemistry sets. Making things that fizzed and went "boom!" had a long-lasting influence on me. But one day of college chemistry class, and I realized I couldn't cut the math. We didn't even get to taste or smell the chemicals—just calculate atomic weights! I did, however, struggle through one semester and made a C. When I realized I didn't have the right stuff to be a real chemist, I switched majors—I chose zoology. I had been diving up and down the Florida Keys since I was fifteen and was fascinated by everything that swam or crawled. I really wanted marine biology, but at that time it was not a major for undergraduates. You first had to earn a bachelor's degree in biology or zoology. So, I majored in zoology and minored in botany with the general plan of eventually going to graduate school.

During my senior year, Patricia and I tied the knot. We had been sweethearts for five years. Pat had been raised in a Catholic family. Because I was a non-functioning Protestant, it was mandatory for me to take "instructions" from a priest in order to be married in a Catholic church. I initially wondered what on earth I could learn about marriage from an unmarried priest! That attitude changed quickly. I learned many things that kept our marriage running smoothly for the next fifty-seven years and counting. The most important single lesson was, when you disagree, "never resort to name-calling." With my knees knocking, we had a fine, full-blown ceremony in a Catholic church attended by many of friends, including my favorite college professor, Dr. Harding Owre. The reader will learn later why I admired her so much.

Spearfishing Days

In those high school and college days—1948 to around 1960—the main reason people went diving was to spearfish. The coral reefs were healthy and beautiful, but few went there just to take in the beauty and you certainly did not go unarmed. In the minds of most, there were sharks and big fish to worry about. Few could afford underwater cameras, and hardly any of those were available in the first place. I was no exception. I carried a spear and became good at holding my breath and spearing fish. We made spears from brake rods—the mechanical brakes in older cars that required long shafts of steel that connected to the back wheels. All we had to do was sharpen one end and add barbs. In the beginning we put notches in the blunt end just like arrows, and we used rubber from old inner tubes that we strapped to a short, hollow bamboo tube. Later, we switched to surgical tubing. The result was called a "Hawaiian sling" but should really be called a "Florida sling." The father of a famous Florida spear fisherman named Art Pinder had invented the "sling." Pinder would eventually become known as "The King of Sling" because of a book written about his life. You shoot the spear shaft about the same way you shoot a bow and arrow. The difference is you are underwater holding your breath and the target is moving.

There were plenty of fish to spear, and spearfishing kept Pat and me from starving during my senior year in college. Why were we so hungry? During my senior year, EK and my mother moved to Japan. It was his FAA job and a chance to change his lifestyle. He wanted to earn more, advance in the system, and satisfy his wanderlust. Pat and I were suddenly on our own. I had to spear a lot of fish for fish markets and restaurants—and play dance-band jobs—to help pay the bills.

To spear enough fish and make some money I had to become a proficient sharpshooter. Fish markets would not buy fish with spear holes in the middle. They had to be speared in the head. That kind of incentive leads to better marksmanship. Proficiency led to competition, so I entered a few local contests. One thing led to another, and in 1958 I competed in, and won, the individual award for spearing the most fish (over two hundred pounds) during a four-hour period at the U.S. National Spearfishing Contest. The contest was held at West End, Bahamas, and based at the old Jack Tar Resort that has long since been demolished. That was where I first met the legendary Art Pinder. My prize was the Johnny Weissmuller trophy, named after the famous swimmer who played Tarzan in the old black-and-white movies.

The following year, our team of three—we called ourselves the Miami Skin Divers—won first place in the National Spearfishing Contest off Miami. The team consisted of Paul Damman, Don Delmonico, and me. Our prize was the giant Owen Churchill Cup and other goodies. Owen Churchill was the inventor and producer of the first commercially available swim fins in the United States. The following year we won third place competing in frigid waters off Laguna Beach, California. Brrrr!

My last competition was in the 1961 World Spearfishing Competition in Almeria, Spain. Our trip was paid for by donations from sportsmen, *Skin Diver* magazine, and Gustav Dalla Valle, the creator of the Healthways Sporting Goods diving department. We beat the Italians, but the Spanish team took first, and the French were second. We were placed third. I will never forget the hair on the back of my neck standing up when we mounted the podium and the band played "The Star-Spangled Banner"! I know ex-

Gene (*left*) and his longtime colleague Harold Hudson pose with their speared fish and six-horsepower Wizard outboard motor in 1954.

Eugene Shinn   **Don Del Monico**   **Paul Damman**

MIAMI SKIN DIVERS CLUB
1959 U. S. A. A. U. National Underwater Spearfishing Champions
Some of their trophies above are: W. J. Voit Memorial Trophy, Owen Churchill Perpetual Trophy, Owen Churchill Cup, Helms Foundation Medallion Awards, Swimaster Trophy and the U. S. Divers Trophy.

Miami Skin Divers Club members Gene (*left*), Don Delmonico, and Paul Damann with Owen Churchill trophy for winning the 1959 United States Amateur Athletic Union (USAAU) National Spearfishing Contest held off Miami. Courtesy of *Skin Diver* magazine.

actly how Olympic champions must feel when they are on the podium at the Olympics and their national anthem is played.

It's difficult now to realize that spearfishing was an Amateur Athletic Union–sponsored event. There was even talk that it might be elevated to an Olympic sport. Imagine even considering the competitive killing of fish an amateur sport today! Yes, I know they chase foxes and kill bulls in the name of sport, but then sometimes the bull gets his revenge. Today, however, there are commercialized spearfishing contests with large money prizes. Competitors mainly use scuba, and private organizations and various sporting-goods and boat manufacturers sponsor the events. The events receive little publicity, but there is a large, underpublicized following.

Spearfishing with various diving buddies helped us keep Miami restaurants and local fish markets supplied with fresh grouper and snapper. It put some change in our pockets and helped us pay our bills. Along the way I met a lot of interesting divers. One of those buddies, named Harold Hudson, I met at the University of Miami. We had both joined the school's "Sea Devils" dive club. Our lives would become intertwined professionally over the next fifty-some-odd years and remain so today.

By then I had several side jobs besides spearfishing and the usual Saturday-night dance-band gigs where I made fifteen dollars each night. My most stable job was making and delivering plastic bags for a small company called Ajax Plastics. I worked there between classes. On weekends Harold and I spearfished, and after selling the fish to a market I would be off to those dance-band gigs on South Miami Beach. I looked good in my band uniform—a white tuxedo—but I often smelled of fish!

Dynamite and Scuba

In the mid-1950s, I hooked up with a fellow named Clarence Dowling. He had worked for my father at the FAA facility at Miami International Airport and had introduced me to spearfishing. Now he had acquired a small salvage boat. With this rig—a converted World War II–vintage landing craft—we would salvage scrap iron from the many turn-of-the-century shipwrecks along the reef line. Clarence had outfitted it with a gasoline-powered lifting crane, and he, or his partner, would operate the winch and crane. My job was diving down and placing dynamite charges in strategic places. Iron plating had to be broken into pieces light enough to be lifted aboard. It was trial and error at first. The iron consisted of inch-thick sheets of riveted plating, and I learned that the strategic place to put dynamite was along rivet lines. The thick plates were all that was left of the steel ships, many of which had been purposely grounded on the reefs for insurance purposes, because cargo ships were switching from coal to more efficient fuel oil.

Our rig could not lift pieces weighing more than one ton, so my job was to make the plates smaller. I had to learn the efficient use of dynamite. The sticks had to be placed end to end along riveted seams. One stick would set off the next one to create a tearing chain reaction.

Hard to believe, but in the 1950s, anyone twenty-one or older could purchase dynamite. It came in fifty-pound cases and cost about seventeen dollars per case. We sometimes blew as much as a case in a single shot! The actual shooting was done from a small dinghy. We used about one hundred feet of ordinary lamp cord and a car battery. I must say it was exciting. After all, the first thing I made with my first Gilbert chemistry set when I was twelve was gunpowder! Dynamite was a lot more fun!

Today's reader would find it odd that no special permits were required to purchase, blast, and pull up scrap iron from old shipwrecks. Even if permits were required, no one asked. A permit agency was not deemed necessary, at least not in the South. That would change around 1960, when someone dynamited a synagogue on Miami Beach. After that, a blaster's permit was required. It involved being fingerprinted, but there was still no special test of proficiency. I guess if you stayed alive, you were proficient.

When working, we seldom saw other boats. Most outboard motors were in the five- to ten-horsepower range, and we were salvaging five miles offshore. That's a long, slow haul for a small outboard. There certainly were no other divers in the vicinity. Very few people ventured to the reefs, and scuba diving was slowly evolving. Those who did venture into the water with spears swam with flippers and a dive mask called a "face plate." Snorkels were coming on line, but we didn't use them. We looked down on those who did. We called divers who used them "lids." That was because the early ones had complicated flaps or lids on top to keep water out. Many used a ping-pong ball in a little cage. Besides, we had all seen the UDT (Underwater Demolition Team) war movies. Navy UDT divers did not use snorkels. Art Pinder didn't use them, so we didn't need them either.

Coral health and fish decline were not an issue, and even sea turtles were not yet protected. The only boats we saw were lobster fishermen putting out or pulling in traps. For floats, they used one-gallon wine jugs, the kind with the finger-hole handle on the neck. The jugs had a stick and small flag attached. Styrofoam floats and plastic line were not yet available. Fishermen used sisal line dipped in oil or tar. What I remember about the glass jugs is how they would shatter and sink when we shot a half case of dynamite. I won't go into detail about the effect on fish. It bothered me then and is unthinkable today.

Our aim was to get the metal to the scrap yards located up the Miami

River. Scrap dealers were paying sixty-five dollars a ton for scrap iron, and we would bring in ten to fourteen tons after a weekend of work. I received 20 percent of the take. That was a lot of money—a lot more than I made in a week at other jobs! The money helped me ignore the dead fish that fed the schools of sharks that were always patrolling down current. Reverberation of the iron boat hull when the lifting-crane engine was running apparently was what kept sharks away. We could always see their fins breaking the surface a few hundred feet down current, but they never came directly under the boat where I was working. At night they would move in and eat all the dead fish. We ate the big fish.

During my junior and senior years I earned extra credit serving as an assistant laboratory instructor. The class was invertebrate zoology, and most students were premed. For the most part, they hated the lab work. The instructor was Dr. Harding Owre. For some reason, she really liked me, probably because I had earlier done an extra-credit science project under her supervision. I was into microscopic plankton, and that was also her specialty.

My lab job was to retrieve the various preserved organisms from containers of formaldehyde and pass them out to the students. I had previously taken the class, so I knew all the critters and their parts. It was understandable that the premed students detested memorizing the names of all the parts on a crab's legs and claws. There was some interest in study of the leaches, tapeworms, and their various segments. They might use that knowledge later in their chosen profession. My knowledge of tapeworms paid off for me many years later.

Shooting in Cuba

Explosives and diving can be exciting, especially far from home. That combination led to an adventurous summer in 1954. I don't remember how it came about, but I was hired by Western Geophysical Company to help with seismic surveying. It was a survey to help find oil—or, more precisely, to find the areas that might have oil. We worked on Cay Sal Bank in the Bahamas and then moved south along the north coast of Cuba. My job was to serve as diver and explosives rigger.

There were two boats. One was the hundred-foot-long shooting boat where I worked, and the other larger one was the recording boat. The recording boat had a twenty-foot-diameter reel mounted on the back deck, which held several hundred feet of three-inch-diameter clear plastic hose. The hose, which contained hundreds of wires, connected to dozens of little black detectors called geophones. Geophones are basically little microphones—often called jugs—spaced at various locations within the hose. The hose was filled with diesel oil to maintain buoyancy and keep saltwater out. This "streamer" of geophones would be trailed behind the boat to pick up reverberations created by explosives that we dropped from the shooting boat. The vibrations that bounced off various layers down deep in the Earth were recorded as data on large phonographic discs. That's right—on phonograph records! To my knowledge, tape recorders for such work had not come into use. The discs looked to be about three feet in diameter. Data were also printed out as what then looked to me like squiggles on graph paper. Occasionally, a seaplane would land to collect data and fly the information back to company headquarters in Houston, Texas.

We started out with thirty thousand pounds of an explosive called nitroamone. Nitroamone, the actual explosive, was a pink powder sealed in three-foot-long metal cylinders. The explosive is basically ammonium nitrate (fertilizer) saturated with something like diesel oil. The stuff was pink and had a pleasant odor. A cap and booster—a charge larger than the conventional dynamite cap—could be fitted into a recessed chamber at one end. One of my jobs was to tape or wire several of the cylinders together and hand them to the master blaster, who worked in a screened-in cage at the aft end of the vessel. The screen prevents stray static charges that might set off an explosion prematurely. That could ruin your day. There were several stories of seismic boats that literally disappeared in a cloud of smoke!

The shooting boat also had a good-sized reel on deck. This reel contained what seemed like a million miles of piano wire. As the boat cruised along, the wire played out through a counter. The counter measured how many feet of wire had come off the reel. With a signal from the man watching the counter, the blaster inserted the blasting cap and booster and pushed the charge overboard. The electrical wire played out as the charge settled back behind the boat. When the boat was about two hundred feet from the charge, the blaster communicated by radio with the recording boat and on a signal pushed the firing button. While piano wire and shooting wire was playing out, I would be preparing the next charge and take it to the blaster in the screened cage. The same process would be repeated many times each day.

The wire measurement ensured that all shots were evenly spaced. Even today one can fly over the Bahama Bank or grassy flats near Belize, in Central America, and see evenly spaced white sand-filled shot holes. The shots usually made a shallow depression about fifty feet in diameter.

In the early 1950s, similar surveys had been conducted off the entire length of the Florida Keys. The shot holes in the Keys were still visible in the mid-1960s, but the grass grew back and the holes were invisible by 1970.

There was a curious structure on the east side of the island of Key Largo near where the original highway (U.S. 1) joined Key Largo. The earthen structure extended several hundred feet offshore. Not many know it was formerly called the dynamite docks. That's where trucks would drive out to load explosives on seismic vessels back when there was active exploration in progress.

What the early workers did and what we were doing were called seismic-refraction surveys as opposed to seismic-reflection surveys. Seismic reflection would later become the more common method. Explosives were eventually replaced by air guns that use high-pressure air to create a shock wave. Air guns are not shot on the bottom and do not leave "shot holes."

On Cay Sal Bank in the Bahamas, we prepared two really big ones. They were five-hundred-pound shots consisting of nitroamone canisters strapped to a wood pallet. As diver, I helped place the pallets on the bottom in about forty feet of water. As expected, the shots made large geysers, just like something from a World War II–era movie.

After surveying Cay Sal Bank, we steamed south to the north coast of Cuba east of Havana. There we worked mainly in lagoons that separate offshore islands from the mainland. The islands were similar in shape and origin to the Florida Keys; both are old Pleistocene coral reefs. The lagoonal environment between the offshore islands and the mainland is comparable to that of Florida Bay, but narrower. My job was to use a skiff and transport the charges to areas too shallow for the shooting boat. Once there, I would get in the shallow water and push charges into the muddy bottom with hands and feet. Some years later, I would find myself pushing core tubes into similar mud bottom in Florida Bay.

One of the benefits of this job, besides good summer wages, was that I saw a lot of pre-Castro Cuba east of Havana. I had already seen much of Havana, and their brothels, when I marched there with the high school band. All in all, it was a good job with interesting people, good food, and every night a comfy place to sleep—a foam pad in the cargo area, on top of about thirty thousand pounds of high explosives!

First Child, Secret Diving, and My First Job

Pat and I became parents of our first son while I was in my senior year of college. Soon another was on the way. Eventually there would be three boys. All three were born at Doctor's Hospital in Coral Gables, right next to the University of Miami campus. The first arrived just a few hours before I took my final exam in scientific German. It was my last semester. Thanks to many nights of study, I made the highest score ever, an A, and I was having a migraine headache during the test. Since my teens, I had been cursed with classic migraines. They followed a predictable pattern, starting with visual impairment and flashing lights followed by intense pain and a drop in my pulse rate to around sixty beats per minute. The pain would localize to the side opposite the flashing-lights side, and I would begin vomiting. While taking the exam, I could only read the paper by turning my head sideways!

After graduating I landed a job as lab technician at the University of Miami Marine Laboratory. The laboratory was located in an old apartment building in Coral Gables, where I worked for about a year before helping the staffs move equipment to its present location on Virginia Key. The laboratory on Virginia Key was simply called the UM Marine Lab, but some years later it would become the Rosenstiel School of Marine and Atmospheric Science.

My boss at the Virginia Key lab was a chemist named Sigmund Miller. Sig, as we called him, worked on a Navy-sponsored project. Military contracts were common and the mainstay of the research effort. The Cold War was on, and the Navy put a lot of money into marine research.

We were to study and, hopefully, discover ways to kill barnacles and other nasty critters that eat boats and pilings. We tested hundreds of compounds in our attempts to find the ultimate killing agent. Fiberglass boat

construction was just beginning, so most were still wood. I learned to impregnate wood test panels with various noxious chemicals using a vacuum
and pressure chamber. Being able to impregnate wood with this method
would eventually lead to an even better job, impregnating marine sediments with liquid plastic. But that came later when I landed a job with a
future.

Secret Diving

While I worked for Sig, another experience was presented. I met underwater photographer Ed Fisher, a close colleague of legendary underwater photographer Jerry Greenburg. Jerry was the first manufacturer of underwater
camera housings on the East Coast. They were called Sea Hawk housings
and made mainly for the Argus C-3 camera, the most popular and afford-
able 35-millimeter camera at that time. I could not afford one.

The way I met Ed Fisher was through photography at the Marine Lab.
He had constructed a plastic housing for his Leica, a German-made 35-mil-
limeter camera that had the best lens available at that time. He used a close-
up lens attachment and built a focusing frame that protruded out about
one foot in front of the lens. Anything within the frame would be in focus.
There was also a flashgun that used flashbulbs—under water!

Using my original handmade fourteen-foot Chris Craft kit boat, we went
about six miles offshore into the Gulf Stream. It was the first time I had
dived in water that seemed bottomless. Ed used scuba and drifted along
with the current about twenty feet down. When various forms of plankton
drifted by, he would position the subject within the framing device and
click the shutter. I would retrieve the used flashbulbs as they bobbed to
the surface. I believe that the photos Ed took were the first of their kind.
In those days, most studies of plankton used those caught in plankton nets
and preserved in formaldehyde. The plankton were usually mangled and
deformed compared to when they were living. The live critters had unexpected long and delicate tendrils and antennae that most plankton experts
had never seen. Photographing plankton this way would become common
in later years, but Ed was the first at most everything he did!

Ed landed a contract with the Navy to photograph experimental airdrop
mines in a designated test area off Fort Lauderdale, Florida. The test site

was in 125 feet of water. The mines had been painted with the experimental antifouling paint that Sig Miller had developed. The secret formulation contained more copper than any other then available and was later marketed under the brand name Dolphinite. Ed would take close-up photos of the key components, and I would take the general views.

Making the 125-foot dives to routinely inspect and photograph mines led to an even more exciting experience. For this, we had to obtain security clearance. The task was to film a Navy destroyer, passing overhead at full bore! The ship had some secret devices mounted on the hull, including rows of ports that released streams of air bubbles. The bubbles tend to disguise a ship's sound signature. Ed shot still photographs, and I operated the Bolex 16-millimeter movie-film camera, which he had taught me how to load and use and which created a passion that would stay with me for life.

To safely photograph the destroyer, another diver working with us named Shale Niskin set up two tall spar buoys held apart by a long aluminum ladder strung horizontally from the buoys thirty feet below the surface. The ship, with its secret device bolted to the bottom, would come roaring overhead between the buoys while we clung to the ladder taking photographs. The feel of the throbbing giant propellers can best be described as awesome. Shale's arrangement worked! We escaped being chewed up in the throbbing screws.

Shale was inventive. He lived aboard a wooden twenty-six-foot sailboat of his own design. It was made of cedar strips, and inside it smelled like a cedar chest. His specialty at the laboratory was designing and making oceanographic research devices. He later formed a company and began designing and selling oceanic research equipment. The Niskin bottle, currently used worldwide by oceanographers, was his best-known invention.

On a warm summer day in 1954 Ed Fisher did something truly revolutionary. Wearing an early quarter-inch foam wet suit, he entered the eighty-four-degree water at French Reef off Key Largo equipped with underwater camping gear, a gas gun, a rubber-powered speargun, several large reserve air tanks, and numerous small devices that included a faulty depth gauge. He was testing a new regulator called a DiveAir, recently designed by Miami colleague Paul Arnold. An automobile inner tube, lashed to surrounding corals, was inflated with air and served as a recliner a few feet off the bottom. His goal was to spend twenty-four hours underwater! At a depth of thirty feet, his wet suit slowly compressed to about the thickness of paper and lost its thermal insulating properties. Nocturnal shark behavior was not yet understood, but he saw only one that darted away. Ed also observed parrot fish sleeping in their nocturnal mucus bubbles. When dawn arrived he was suffering hypothermia, which severely affected his reasoning and thought processes. Nevertheless, Ed speared a snapper for an underwater sushi breakfast and remained underwater until afternoon, fulfilling his vow to remain under for twenty-four hours.

The local press treated the event like a flagpole-sitting stunt. Ed responded, "Someday man will want to spend even longer periods of time under water." We now spend millions of dollars on dry underwater habitats so divers can live submerged and swim outside to conduct research. Ed was and remains a true visionary.

Tipping Point and a Fork in the Road

My day job at the University of Miami Marine Laboratory paid a whopping three thousand dollars a year! If it were not for all the lobster, crab, and fish we could eat, as well as the occasional dance-band gigs, the Shinns—now numbering five—might have starved. We were so fed up with fish and lobster that we tried everything imaginable to disguise its flavor. Nothing we did could make seafood taste like steak. I was reaching a turning point and another fork in the road of life.

My tipping point came the day Sig asked me to help a union plumber install pipes under the new chemistry laboratory at Virginia Key. As we crawled together in the dirty and spider-infested dark space under the first floor, I learned that the union plumber made more than twice my salary. I started thinking more about graduate school or maybe becoming a plumber!

This time down life's road I took the wrong fork, but fortunately it was short. I went through the graduate school application process. I had been told by one of my advisers that without more algebra I could never be a scientist. I would not be allowed to take any graduate marine biology classes until I had first finished the math classes. There was also the daunting prospect of passing another German exam. I had squeezed through two years of scientific German as an undergraduate, but I had forgotten a lot and I feared failing and did not look forward to relearning German.

Students today are fortunate in one way. Foreign language is no longer required for a PhD. They can take computer-programming courses instead. In spite of my concerns, I agreed to the stipulations and signed up for yet another algebra class. I had already taken two years of algebra as an undergraduate and slipped around others, including calculus, by taking courses in logic, philosophy, and science history. After attending two algebra classes (I had been out of school for a year), it was just more déjà vu than I could stand! This was during the beginning of the "new math" era. We were not to just learn how to do algebraic equations but were supposed to understand it as well. Thus, we proved $2 + 2 = 4$. I could see no practicality in it. I had learned more useful math in the chemistry class through which I had struggled. I quit graduate school after one week and was in the process of applying to be a grade school teacher when Sig Miller said, "I have this friend who runs a small laboratory for Shell Oil Company, and he is looking for a lab technician." His lab tech had had a life-altering diving accident, and his job might be available.

2

The Shell Years

Start of a Profession

This time I took the right fork. I went to meet Sig's friend. His name was Dr. Robert N. Ginsburg. Ginsburg was running a small research laboratory located in an apartment complex in Coral Gables, Florida. When I arrived there were three people in the lab including Ginsburg. The others were Ken Stockman and John McCallum. One of the original members of the group, Michael Lloyd, had returned to school for a PhD. He was at Cal Tech studying isotope geochemistry under Heinz Lowenstam. Lowenstam had been Ginsburg's major professor when he was a student at the University of Chicago. I would meet Mike later, and he would become a trusted lifelong mentor.

The company was Shell Development Company, the new research division of Shell Oil Company. "Shell D" had been set up to do research for the parent company in Holland after World War II. The company, with headquarters in Houston, had decided to support a new and innovative line of basic research. Management thought this new approach might provide a slight competitive edge in the search for oil. It was based on the theory and backbone of geology called *uniformitarianism*, which I'd never heard of. This paradigm of geology basically says, "The present is a key to the past."

Nearly all oil and gas are produced from sedimentary rocks, and about half the world's sedimentary rocks consist of quartz sand known as "clastics." Quartz sand turns into sandstone—it's also the stuff of glass bottles. The other half is limestone. Limestone consists of shell fragments, corals, and many other particles made of calcium carbonate. Sometimes they are mixed together. Geologists who work on nonsedimentary rocks, such as granite and hard minerals—we call them hard rocks—call limestone and sandstone "soft rocks." Hard-rock geologists tend to look down on soft-rock geologists.

When sedimentary grains were forming in the past, before they became rocks, they were affected by the same physical processes of sedimentation that occur today. In short, water has always flowed downhill! That's uniformitarianism in a nutshell. In other words, if you understand how and where the rocks/sediments are forming today, you should be able to better understand how and where it happened in the past. If you knew how the rocks formed then, you might be able to predict where rocks with lots of holes might be—and there in the holes, we may find oil. That's the big jump! Some might call it a leap of faith, but it's logical.

Bob Ginsburg's mission was to study how limestone forms. At the same time, other researchers at the home laboratory in Houston concerned themselves with quartz sand and how it becomes sandstone. Some very venturous geologists and forward-looking management had decided this kind of knowledge might provide Shell and its petroleum geologists with an edge over the competition. The search for oil and gas was, and remains, a very competitive and expensive enterprise. I soon learned we were way out in front of the herd. Best of all I was learning something new. I was, however, missing a lot of basic facts one learns in college. I would spend the rest of my life learning what I had missed.

Ginsburg had just earned his PhD at the University of Chicago studying carbonate sediments in the Florida Keys. He had studied under the GI Bill, which funded students who had served in the military. Bob had served in the Philippines. He also had lots of stories about seasickness aboard troop carriers.

Bob's professor at the University of Chicago, Heinz Lowenstam, had also taught one of his classmates, Cesare Emiliani, along with George Edwards—two very famous isotope geochemists. The list of famous graduates from the University of Chicago is long. I would later learn that this group of classmates would dominate their fields and were jokingly called "The Chicago Mafia." The University of Chicago, in addition to its role in the Manhattan Project, had spawned outstanding scientists in many fields.

There were several reasons Shell Oil Company sponsored this little group in Coral Gables. First, Bob had just completed his dissertation on carbonate sediments in the Florida Keys. Another was a unique technique Ginsburg and his colleagues had developed for taking push cores of modern sediment. Taking cores was not new, but these sediment cores were then impregnated with clear polyester resin. The plastic resin, a version of that used to make fiberglass boats, hardens in a few hours. The plastic-impregnated core could then be cut with a diamond rock saw and, when sliced, looks almost like real limestone. The technique allowed geologists to "skip" the thousands of years it takes for sediment to become rock. These modern sediments could be compared with ancient limestone millions of years old. It really made me believe in uniformitarianism. By this time, not only could I pronounce that long word—I knew what it meant! Part of my job working for Sig Miller had been to vacuum-impregnate wood samples with various creosotes and pesticides. Thus, I already knew how to impregnate samples and I was mechanically inclined.

Geologists working in the "oil patch"—the name for those actually working with oil and gas—loved the technique. They could see so many similarities with the rock cores and chips brought up from formations deep in the Earth—especially those rocks that produced oil! We were convinced this approach would help the oil-patch guys find the next multimillion-dollar oil field. And it did.

In retrospect I can say, as others have, that the postwar years from 1950 up to the early 1970s were the golden years for research. Money was available, researchers were energetic and innovative, and bureaucracy was almost nonexistent compared to today. Paperwork was minimal, and every project promised great advances. It seems more paperwork came with the computer that sped up communications—of information that was not always necessary.

It was September 1959. My starting salary was $4,800 per year—I was ecstatic! Not only that, I would be diving in parts of Florida Bay that reminded me of the lagoon in Cuba. This time it would be taking cores, not planting explosives! Even better, I would be collecting biological samples and sediments on the Florida Keys reef tract that I already knew so well. This was Heaven on Earth. My biology background would pay off, as most lime sediments, including much of the Earth's limestone, consist of the remains of calcareous organisms—critters! Even better, the growing Shinn family was no longer dependent on fish and lobster. Steak really tasted good!

Fortune or Misfortune

Most of us look back on our careers and think, what if? Where would I be if such and such hadn't happened? Did I take the right fork in the road? I think about that often. In my case, it sadly came as a result of another's misfortune. Bob Ginsburg's previous lab tech, a fellow named Reynolds Moody, had suffered a crippling diving accident.

Deep diving with scuba was Reynolds's hobby. He did his deep diving on weekends. One day while diving, he pushed it a little too far. The limits of deep diving were not well known in the 1950s. Reynolds was forty-eight years old. He told me that once, as a marine pilot, he briefly held the world's altitude record. Now he was going the other way. He was routinely making bounce dives to more than two hundred feet with a single tank of air.

It was generally thought, and I relied on it for years, that with a single tank you could not go deep enough, long enough, to get the bends. For most healthy young people, that still pretty much holds true. By the time you reach 200 or 250 feet, and if you're not overcome with nitrogen narcosis, there is so little air left in a single tank that you are automatically forced to surface before your tissues saturate with nitrogen. It's the nitrogen that causes the "bends." The kind of double-hose regulators available during those early days of diving provided plenty of warning when air was running low. First, it became noticeably difficult to breathe, but as you ascended, breathing became easier. A slow ascent would get you safely to the surface. Ascent included a ten-minute stay about fifteen feet below the surface.

On that fateful dive, Reynolds apparently came up too fast and experienced an air embolism, a bubble in his bloodstream. He made it to the boat, climbed aboard, and passed out. His dive buddy hailed a passing freighter, and he was hauled aboard. After a radio call, a Navy helicopter picked him off the deck and transported him to the only recompression chamber in south Florida, which was at the Navy underwater diving school in Key West. Back then none of us knew that you do not fly the same day you dive. Flying increases the chances of decompression sickness.

Reynolds remembered that when he awoke in the chamber he felt fine. However, when the pressure was relieved a few hours later, a bubble had moved and had lodged somewhere in his spinal cord, cutting off blood supply. He was paralyzed from the waist down. Because of his accident, I had the opportunity to begin learning sedimentary geology. It is a sobering thought that still haunts me. The rest of my life story would have been very different if Reynolds had not suffered the accident.

Reynolds and I became good friends, and I frequently visited him in the afternoon on the way home. During one visit he told me the story of how he had lost a close friend, a diver named Hope Root. Hope had attempted to set the world's depth record using scuba. In the Gulf Stream off Miami, he was last seen as a trace on the Fathometer at about six hundred feet. I decided never to try setting depth records. Over the years, I have known others who died trying.

Oil-Patch Visitors

One of the duties of Bob Ginsburg's four-person lab was to host geologists from the "oil patch," slang for geologists actually working out where the oil is produced. They came, sometimes two at a time, for six-month training assignments. During their stay, many conducted field studies and learned firsthand how calcareous sediments are created. They studied everything from mud to coarse-sand particles, and each person usually conducted a small project involving field studies. For these activities I was guide, boat driver, diver, sample collector, and the sample-preparation guy. It was like a field course in sedimentary geology, and I was literally learning from the bottom up. I did not realize at the time that we were making discoveries and observations long before they reached geology textbooks. It was truly a bootstrap beginning for me. With my biological background, I knew the critters that created the sediment, including the ones that scratched, dug, and blended various sediment grains together. I cored the sediment, converted it to rock with plastic—the technique that Reynolds and Bob had developed. I sliced and examined the "instant limestone" before handing the cores off to the geologists doing the real studies. Every core slice revealed a new question, and I was curious and eager.

All of those six-month assignees became my teachers. The first was Real

Turmel. He would also be my teacher a few years later when I too went to the oil patch. Another was a fellow named Pete Rose. He had been using fossil foraminifera (a shelled protozoan) to unravel a geological formation in Texas called the Edwards Limestone, where Shell was looking for oil. Pete was collecting live foraminifera from different environments to determine where they did and didn't live. We did a lot of diving and sampling together. Our lives would cross many years later, and he would set me off on a new road.

The next geologist from the oil patch would be Duff Kerr. Duff was involved in the aftermath and sedimentary modification brought on by Hurricane Donna in 1960. He would also be one of my teachers when I visited the oil patch in Midland, Texas.

Another six-month trainee was Jim Rodgers. Jim worked on a tidal delta in the Keys just off Whale Harbor. We took many sediment cores and mapped sediment thicknesses using a metal probe. The water was usually murky, and Jim had a habit of diving down and coming eye to eye with sharks. Jim eventually left Shell and discovered his own oil—ironically using some ideas he learned from our published work on the shape of meandering tidal channels on the Andros Island tidal flats in the Bahamas.

There was also Don Baars. Don too worked on tidal deltas. He had been a string bass player in an earlier life, so we had a musical connection. Don left Shell rather early. His specialty was subsurface geology in the Four Corners region where Texas, New Mexico, Kansas, and Colorado meet. Later in life he ran river trips for geologists and published a book about his adventures.

And then there was one of my favorite colleagues, Dr. Mahlon Ball, and his wife, Marilyn. Mahlon was a Kansas boy who had been an Underwater Demolition Team diver in the Navy. We both did a lot of diving together and had many adventures—sometimes when we were inebriated, such as the night in Bimini when he pushed me onstage to play drums with the Sparrow (a well-known Calypso band at that time) or surviving midnight storms on the Great Bahama Bank. Two of Mahlon's favorite expressions were, "It's a good life if you don't weaken" and "The world's gone mad." I adopted those expressions, and they apply more today than when I first heard them as a twenty-year-old kid.

Another fellow, an Australian, came for training. He was interested in

many aspects of geology, including my little weekend coral-growth experiments. His name was Bob Foster. We would find ourselves working in
Doha, Qatar, in the Persian Gulf a few years later. Some years after that, he
would bring me to testify about my coral work before the Australian Great
Barrier Reef Commission. You can't always predict the future or where
forks in the road will lead. It's important to take the right fork!

Learning to Write about Science

Bob Ginsburg and his wife, Helen Sloan Ginsburg, more or less adopted the
expanding Shinn family. Bob noted my increasing interest in geology and
natural history and fed that appetite with the classical geological literature
as well as current geological theories. He was always asking questions. I was
especially drawn to the idea of "multiple working hypotheses," an approach
basic to geological science. A big question at the time was, how does the
mineral dolomite form?

Helen taught creative writing at the University of Miami. Naturally, I
took her writing course and learned many valuable lessons. Unfortunately,
she could not teach me to spell. It was too late for that. Interestingly, I had
little problem spelling long German words when I was learning to read
German. It's a logical language, but English was another matter.

Both Helen and Bob stressed, "When you write, you need a point of
view." Just writing about the research you did is not enough. What did the
results mean? What was the question being asked? I adopted their attitude
that implications revealed by the research should be the main point of a
scientific paper. There was another philosophy usually not learned or em-
phasized in school. It had much to do with why we did things.

Because we were working for a petroleum exploration company trying
to get an edge on the competition, *ideas* were considered more important
than reams of data. In the oil patch, data are limited. There is never as much
information as you would like. Many data can come from a single well, but
wells are rarely numerous. Someone has to decide where to drill the next
million-dollar test well before the competitor does.

Contouring a map using subsurface data from a limited number of existing wells is where creativity can make the difference between success and
failure. Everyone in the business usually has access to the same data. It's the

interpretation that makes the difference. Piles of data can bog you down while the competition moves ahead and drills a well. There is pressure.

Of course, I didn't realize we were on the cutting edge of geological science at the time. How could I have known? What we realized was that if the geologist in the oil patch could inspect a core sample from five thousand feet down and recognize the environment in which it had formed, he or she might then have an advantage over the competition. So if the geologist could recognize from a rock core that the limestone had formed on, let's say, a tidal flat, then knowledge of modern tidal flats could provide a hint as to what shape the reservoir might have had. It would be reflected in the way the contours were drawn.

For example, think of a sinuous meandering channel filled with porous sand. That envisioned pattern could lead to contouring a map differently, assuming you realized it was a channel deposit in the first place—channels have porous sand that could hold oil! If, on the other hand, the rocks looked like a reef, then a different shape should be expected. All of this information and the patterns stored in the brain can influence how contours are drawn on a map differently from those that someone else might draw. I venture to say that drawing contour maps by the numbers, the way computers do, will not provide that intuitive advantage. After all, everyone has the same data set, and computers without knowledge of what the rock means will draw pretty much the same map as all the other computers. Ideas—some call it intuition—remain the most important tool in oil exploration. Remember Spindle Top? At the turn of the nineteenth century, every classically trained geologist knew there was no oil associated with salt domes. Nevertheless, Anthony Lucas drilled a salt dome called "Spindle Top" near Beaumont, Texas, in 1901, and the gusher it produced brought in a new era of oil exploration around the Gulf of Mexico. Structures associated with salt domes remain the major targets in the Gulf area, especially in the deep gulf now attainable via semisubmersible drill platforms.

This philosophy of ideas has stayed with me to this day, and besides, the reader now knows I was never good with numbers. Those who love numbers often get hung up in them! As my friend Mahlon Ball used to say, "Gene, if you have to use a lot of statistics, go find another problem." Another one of the visiting geologists I worked with used to say, "Figures don't lie, but liars can sure figure." I developed a philosophy that if I can photograph it, I know it's real. One roguish friend, another six-month trainee, once called me a "sediphotographer" because I was always photographing sediments and sedimentary structures.

All this discussion explains why we did not publish great piles of data. As Bob Ginsburg used to say, "There are hunters and there are cooks. In a primitive society, the hunters bring home the game, the cooks prepare it." We were hunters! We hunted in the field for new ideas and then published idea-driven papers. Cooks could cook up the data and write the textbooks later. We were into discovery and exploration.

Of course we did collect data, but we did so for proof and hypothesis testing. A paper that is all data is not very interesting, and important points and implications might be lost on the reader. It was all related to what has become Ginsburg's famous "SO WHAT?" question. Maybe he didn't invent the concept or the question, but he sure used it on us a lot, and legions of students attribute that admonition to Bob.

Independent Research

The research bug had bitten me. That professor back at the University of Miami Marine Lab who told me I would never be a scientist still haunted me, and the memory became a challenge. I had to prove him wrong. One of the advantages I had during those early years was that I had a boat. The first I had built from a kit while still in high school. It was the fourteen-foot Chris Craft kit boat powered by a six-horsepower Wizard. It kept the family in fish, and anything I could scrounge off the bottom. It got me through college. With my new salary at Shell, I moved up to a used sixteen-footer powered by a secondhand twenty-five-horse Johnson. The outboard motor forced me to become a good outboard mechanic to avoid swimming home, which I did have to do once! During this period of life, our kitchen table saw many disassembled outboard motors. Sometimes, VW engine parts demanded the table. A Volkswagen "bug" struggled, but it towed my boat and trailer up and down the Florida Keys for several years. Pulling boat trailers burned exhaust valves. I repaired many engines.

On weekends I could get out to the reefs, weather permitting, and still did some spearfishing. This was a period when many of us began competing in breath-holding spearfishing contests. Even Pat competed after having her third child. She won the Southeast Women's Spearfishing Division in 1960. After my 1961 experience at the World Spearfishing Tournament in Spain, however, I hung up the spear. We were starting to notice the decline of fish and reluctantly admitting we might have something to do with it. Besides, the Shinn family could now afford steak!

Early on, Bob Ginsburg had taken me aside and said, "Gene you are seeing many things out on the water that few scientists see. You should do something more meaningful than spearfish." I took it seriously and wondered what mysteries I might solve. It didn't take long to come up with an idea. It would involve two things I could do well—dive and use dynamite!

One of the obvious reef features all divers and geologists had observed were the massive finger-like projections on the seaward side of coral reefs. We called them "spurs and grooves." What made them? Why were they always perpendicular to the main reef? No one seemed to know, yet they were the most obvious topographic feature on reefs the world over. Some geologists had published papers stating that they were constructed by coral growth. Others said they were created by erosion. If only one could look inside, there might be some hidden clues. We hadn't yet developed a simple way to core and peer inside. Our coring was restricted to pushing tubes into soft sediment—the softer the better!

Why were we restricting our work to areas that could be push-cored? One of the guiding paradigms, based on observations of ancient limestone, was that most of the limestone in the geologic record was made of rocks that had once been soft lime mud. Even when we looked at the classical ancient reefs, whether they were four hundred million or only ten million years old, we usually saw fossils surrounded by what was once fine-grained lime mud (or so it was thought). We had coral reefs off the Florida Keys, but because of the rule, we thought they were not like the ancient reefs. We avoided them. Besides, we didn't have the drilling techniques to even attempt their study. We couldn't see what was inside at that time, at least not officially.

There were, however, inshore mud banks with corals, such as at Rodriguez Key bank and Tavernier Key bank. We could push cores into them by hand and impregnate the cored sediment with plastic. When we sawed the cores, they looked exactly like some ancient reefs hundreds of millions of years old. The finger corals that grow on these banks are seen to be floating in lime mud or silt. For that reason we saw no reason to study the big beautiful offshore reefs, which, to us, were simply pretty places with lots of multicolored fish. Admittedly, diving on coral reefs was much more fun, but officially we restricted most of our research to those muddy areas where the water was usually turbid and the pretty fish, if there, could not be seen.

Underwater "Explosures"

In spite of the scientific rationale against studying coral reefs, I still found the outer reefs, and especially their spurs and grooves, interesting and wanted to know how they had formed. I had a boat, and I still had access to dynamite (I was over twenty-one). What I did with it on weekends had no official connection with Shell Development Company. So, clearly, there was only one logical way to see what had made the spurs. Why not blow some up and see what's inside? Geologists study outcrops—places where rocks are exposed—so why not make an underwater outcrop? Call them "explosures"!

Surprisingly—and with explosives—I found that spur interiors contained coral skeletons different from those growing on the surface, and there was a lot of mud within them! The spurs had been built by *Acropora palmata*, a species commonly called elkhorn coral. The surface, however, was crusted over with *Millepora* (a fire coral). Only a few scattered elkhorn corals populated the surface. There was muddy sediment inside, not the corn-flake-size grains that littered the surface and filled the grooves separating the spurs. The significance of that observation I would not appreciate until many years later.

In a few areas such as at Grecian Rocks—the reef was called Key Largo Dry Rocks back then—it was easy to observe that elongated and oriented branches of fast-growing elkhorn were creating the spurs. Such places were rare. Regrettably, most all of that elkhorn began to die in the late 1970s and early 1980s. Little is living today. That coral species also began dying throughout the Caribbean at about the same time it died in Florida.[1]

With Bob Ginsburg's encouragement, I prepared my first scientific paper, "Spur and Groove Formation in the Florida Keys," which was published in the *Journal of Sedimentary Petrology*. The major point was that the Florida spurs and grooves were created by coral growth rather than erosion. Those in the Pacific were thought to be the result of erosion, something I was to appreciate firsthand many years later.

My first paper was a great success, probably because I had the field almost all to myself, and I had used a unique study technique—dynamite. Few cared about the environmental effects of dynamite then, as salvagers had been using it up and down the reefs for many years. Concern would come with widespread use of scuba and an exponential increase in the number of diver-related impacts.

After that first paper, I was especially pleased to receive a congratulatory letter from Francis P. Shepard, a famous marine geologist at the Scripps Institution of Oceanography in San Diego. Today, there is a special marine sedimentology award given annually in his name.

Management at Shell was a little concerned about my sampling methods, but they had little control over my weekend activities. Nevertheless, I did stop using that technique in the Florida Keys. Times were changing. After a synagogue on Miami Beach was dynamited, two FBI agents paid me a visit. Of course I was not guilty, but I decided it was probably not a good idea to keep the stuff in my laundry room anymore.

Bob Ginsburg saw the value of this technique, and years later he literally took the technique to greater depths. Using the two-man submarine *Nekton*, he created "explosures" several hundred feet down on reef walls, well below the zone of active coral growth. As with the shallow reefs, explosives were the only way to get an adequate sample at seven hundred feet. Ginsburg, working with Noel James, while using the *Nekton* submarine piloted by Richard Slater, made many new discoveries, and benchmark papers came from the use of explosives in depths beyond those accessible with scuba.

One of the unappreciated advantages of working at the Coral Gables Shell laboratory was that just about everyone in the field of carbonate sedimentology would eventually visit Bob and his four-person team. I didn't know how important these people were then. They themselves probably didn't know either.

It was at about this time that Edward Hoffmeister made a visit. He had come from the University of Rochester in upstate New York and had taken a visiting scientist position at the UM Marine Lab on Virginia Key. Ed had been one of the early reef workers. Ed, Josh Tracey, and Harry Ladd had worked together at Bikini and Enewetak Atolls in the Pacific during the early days of nuclear testing. They were well-known and highly respected reef geologists.

Hoffmeister had been a disciple of the famous geologist T. Wayland Vaughan. In the early 1900s, Vaughan had studied the geology of Florida and the Bahamas. He also studied corals and their growth rates at the Carnegie Institute's Laboratory at the Dry Tortugas in the Gulf of Mexico sixty miles west of Key West. Vaughan later went to California to direct the Scripps Institution of Oceanography.

About the time Ed came to Miami, Gray Multer also arrived. As a team, they began to investigate the geology of coral reef limestone—the rock—that is the Florida Keys. To help understand these 125,000-year-old corals, they felt it necessary to know how fast corals grew in the Florida Keys. Alfred G. Mayor and Vaughan had initiated such studies during the early days at Dry Tortugas. Mayor had conceived of and initiated the famous Carnegie Research Laboratory at the Dry Tortugas in 1905, and Ed had worked under Vaughan and Mayor.

In Mayor's day, diving equipment was limited to bulky suits and diving helmets. But now Ed Hoffmeister and Gray Multer had something different. They had me—and my recently acquired sixteen-foot boat! I soon became their diver/guide.

As Vaughan had done fifty years earlier at the Tortugas Carnegie Laboratory, we harvested small live corals and cemented them to heavy cement tiles. The tiles were placed in ten feet of water on the leeward side of the 110-year-old lighthouse at Carysfort Reef. Every few months, we would go out for their biological "checkup." I would retrieve the tiles, and Ed and Gray would make measurements and take photographs. I was learning and was completely captivated by the results. I wanted to do more! With their encouragement, I began yet another weekend experiment of my own. No dynamite this time! I wanted to know how fast *Acropora cervicornis*, or staghorn coral, could grow. I also wanted to learn why this species is restricted to the outer reefs far from shore.

One weekend I placed plastic bands around the coral branches. The little rings of plastic snapped into place, and each band was placed ten centimeters from the growing tip. I would subsequently measure the distance from the band to the tip. The main site was at what we used to call Key Largo Dry

Rocks (later changed to Grecian Rocks). My study was different because it was all done underwater. All previous studies required removing specimens from the water for measurement. In a few months, the fast-growing coral had actually grown over the plastic bands!

The other part of the study involved transporting clumps from the banded parent colony to sites closer to shore. I did briefly take these transplants out of the water to put them in seawater-filled containers. One of the sites was near shore where *A. cervicornis* had not been seen before—not even old dead specimens. I just wanted to know why this coral had not grown there. There were two more in-between sites located farther offshore in places where staghorn coral was rare. I had also acquired some simple maximum/minimum thermometers like those used by gardeners. The instruments are mercury thermometers with a metal indicator that rides on top of the mercury column. When the mercury moves, the metal tab is pushed along, but it stays in place after the mercury drops back down. The tab shows the maximum and minimum positions to which the mercury had moved. I encased the thermometers in a plastic tube and fixed it to the seafloor at each transplant site. The drawback was that the thermometers only told me what the high and low temperatures had been since the last visit, not what had transpired during the intervening time.

Once again, this little study had nothing to do with our mission at the Shell Development Research Lab. The distribution of this species was known, and we knew it did not live near shore in the Keys, so, geologically speaking, the experiment would not be very significant. I didn't care. I just wanted to know how fast it grew and why it didn't live near shore. The irony, or serendipity, of this interest-driven exercise would become apparent in the future. In 1972, I would be sent to Australia to testify before the Great Barrier Reef Commission. I was there to tell them about the rapidity of staghorn growth in Florida and how it recovers if broken. Minor experiments had shown me that this coral could be broken into small pieces, and the broken branches would quickly heal and form new colonies and new branches. Hurricane damage had already shown that this coral species recovered even when it was mowed down by storm waves. Unhappily, all that changed in the early 1980s, when diseases attacked this and other species throughout the Caribbean Basin.

What I liked about this study was that I could go out on the reef, spear a few fish for dinner, and on the way in stop at each site, record the temperature, and take measurements with a mechanics micrometer. Of course, if you do things like this, other interesting things will usually come to light. For example, the coral at the Grecian Rocks site that supplied the transplants grew at a rapid rate but fluctuated slightly with seasonal temperature change. Because I looked at the same colonies each month, I also noticed that they initiated branching during the winter.

Surprisingly, the colony nearest shore, where *A. cervicornis* was not supposed to grow, grew just as fast as those at the offshore site in spite of the mud and poor water visibility. However, when the temperature in the shallow water exceeded that of the offshore site, the coral lost its color and turned snow white! Years later, in the 1980s, the phenomenon would acquire a name—it became known as "coral bleaching." Surprisingly, this particular coral did not die, and as soon as the water temperature decreased, the colony regained its natural color and resumed growth. As long as the temperature was within the same range as in the offshore waters, the colony actually grew at about the same rate. However, a strong cold front in February chilled the water down to 13°C. The colony died! I then knew why this species is restricted to the outer reef areas near the warm Gulf Stream.

What was also exciting to me was that my measured growth rate at Grecian Rocks was twice what had been found in the studies at the Carnegie Lab in the Dry Tortugas. Another observation, the implications of which would become more important later, was that most of the new branches were initiated during the winter, usually around February.

A Lesson to Remember

While doing this work, I took Tom Goreau, the founder and director of the Discovery Bay Marine Lab in Jamaica, out on the company boat to see the Florida reefs. We had with us an English geologist named Robin Bathurst. Neither had seen a Florida reef. Tom was quick to point out that our reefs were poor examples compared to those in Jamaica. He was impressed with the greenness of the water compared to the clear blue oceanic waters off the north coast of Jamaica. He took a keen interest in my growth-rate study and encouraged me to write up the results. Later he taught me another important lesson that young scientists should heed.

I did as Goreau suggested and wrote up the results for publication. The paper was quickly rejected by a well-known biological journal. One of the editors scolded me for not providing enough data! Not enough numbers and statistics. He said I reached too far with my conclusions based on limited data. When I told Goreau, he said, "Don't change a word. Send it to another journal." So I sent it to the *Journal of Paleontology* and they accepted it immediately. In fact, it was the lead article in the next issue! I forever thanked Tom for his advice. I still make that point to young scientists who have papers rejected. Unless it is a really bad paper—sometimes difficult to determine—try another journal. You learn through experience and rejection. There are fads in science that come and go, and some journals have one paradigm in mind while others may have another. Some like abundant data and statistics while others thrive on new ideas.

Another lesson I learned should be passed on here, and this one I learned from my father. I was writing up results of my work on tidal flats and having a terrible time. No one seemed to like my writing. My reports were especially not well received by Bob's manager back in Houston. I commiserated with my father about it because I knew he was a good writer. EK said, "Here is a little trick that from experience I know will work." He suggested I read some papers written by the manager in Houston, Don Higgs. Bob and I both got along great with Don, but we couldn't keep up with him when it came to martini consumption. So I took my dad's advice and read two of Higgs's published papers. They were about rock mechanics, but that did not matter. The trick was to make a list of words and expressions he commonly used—words that I would not have thought of. So I made a list of twenty-six words. I called them Higgisms.

The next trick was to go through what I had written and find places where my words could be substituted with words from the list. It was actually quite easy. Soon thereafter, Don called Bob and said, "Gene's writing sure has improved." It works! And best of all, I had improved my vocabulary. I still use those words. It was all for the better. Many years later, I used that same trick when dealing with managers at headquarters of the USGS. New gobbledygook "governmentise" was evolving all the time and still is. There were new words and expressions like *stakeholders*, *decision-support system*, *centerpiece*, and other classic jargon. Once again, the trick worked! Take note, you young readers who may be having problems communicating your scientific results.

I had been a laboratory technician for three years and had loved every minute of it. There was the coral work, and there were other things that gave me encouragement. Plastic shrimp! Dolomite! In the meantime, management, with Bob Ginsburg's encouragement, had decided I should be elevated to the position of geologist. It was heaven! I got a raise, and the Shinns could afford even more steak.

Plastic Shrimp

In my lab tech job I worked almost every day with polyester resin—the plastic we used to impregnate cores. It had occurred to me that instead of bringing cores to the lab and impregnating them with resin I might be able to do it all in the field. It is very difficult to push core in sand, especially beach sand. My idea was to pour the resin in a crab burrow. The resin should seep out in the sand and, when hardened, I'd dig it out. Beach-sand laminations should be preserved in the hardened mass around the burrow. I knew where there was a beach close by with numerous crab burrows, just off the Rickenbacker Causeway that leads to Virginia Key. Yes, the causeway is named after the same Eddie Rickenbacker who created Eastern Airlines and whom my father tangled with.

I poured the resin in the burrow and had some left over. It was going to harden in a few minutes, so I stepped out in a foot of water where the bottom was mostly mud and dumped it. Shame on me! The resin is heavier than water, so it sank to the bottom. As I watched, it began to disappear down some small holes made by some kind of burrowing organism. About an hour later, I reached down into the mud and to my amazement removed a perfect cast of a crustacean burrow. It was a total surprise! My attempt to impregnate the beach sand did not work very well, but that didn't matter—this was much better!

Triumphantly, I returned to the office with the little burrow cast not knowing what the geologists would think. Everyone was amazed and excited, but I wasn't completely sure why. The next day I was back at the beach with more resin and made some amazing burrow casts. The resin did not penetrate the mud but remained in the burrows. The casts were smooth on the bottom and rough on the top. The critter that made them was a mud shrimp, also called a snapping shrimp. They can snap their large claw and stun small prey just from the loud shock wave. Soon we began to wonder about the shapes of other burrows, and we now had this new technique to find out more!

There are millions of strange volcano-like mounds of sediment that characterize the seafloor everywhere in the Keys and the Bahamas. We had always thought they were worm or sea-cucumber (*holothurian*) mounds. This time, the resin-cast technique revealed something totally new. These burrows were smooth and deep and they had most unusual shapes. A small burrowing shrimp, called *Callianassa*, was the builder. We knew because we caught some in the resin. This shrimp, also called a ghost shrimp, was already known from the geologic record. Their ancestors and their burrows were preserved in rocks dating back at least one hundred million years!

We would learn more, but it was about this time that management decided we should keep this discovery a company secret. The reason? There is an enigmatic structure called *Stromatactis* in some ancient reefs, and some of those reefs are oil reservoirs. Paleontologists speculated for years over the origin of *Stromatactis*. Did it have something to do with the making of the reefs? Was it a reef builder similar to corals? If *Stromatactis* were, in fact, ancient shrimp burrows, then the structure had once been an open hole. If the burrow somehow stayed open, it would add to the volume of oil such a reef might hold. Now I knew why the geologists had been so excited about the ghost-shrimp burrows. A few years later we decided these burrows were not the answer, but in the meantime it was a company secret.

One day, geologist Robin Bathurst from England visited Bob's office. Bathurst, who later published one of the most famous textbooks on limestone formation, was pursuing the origin of *Stromatactis*. He had concluded that *Stromatactis* was made by some kind of organic structure that decayed after it built reef mounds. These structures always had a characteristic infilling of laminated lime mud overlain by a kind of clear fibrous calcite. At that time, we still thought the burrows we were examining were exactly what he was predicting.

I cranked up the company boat and took Bathurst in the field to look at mud banks and coral reefs. (The same trip with Tom Goreau, described earlier.) That first meeting with Bathurst was painful, as I could not tell him about the mud shrimp and the *Callianassa* we still thought had created ancient *Stromatactis*. Much later and with Shell's permission, I published a paper on the burrows. Photographs of the various kinds of plastic casts we had made illustrated the paper. The paper received the annual best paper award from the Society of Economic Paleontologists and Mineralogists (SEPM). It was my first!

Years later, Bathurst and I could talk about the burrow casts, but by then we had both decided that *Stromatactis* was the product of a much different process. That process was something that would be revealed in the future while I was swimming in the Persian Gulf. Meanwhile, we made many more shrimp-burrow casts—on one occasion it was while on board Captain Jacques Cousteau's ship, the *Calypso*. That cast was about six feet long. Cousteau made similar casts in one of his later television shows. By then, he had taken credit for inventing the technique.

Dolomite and Tidal Flats

One of the more significant areas of research was the vast expanse of tidal flats on the west side of Andros Island in the Bahamas. Earlier, Bob's team had worked mainly on the shallow Great Bahama Bank. During a company field trip, Shell geologist Perry Roehl was standing on the western shore of Andros Island discussing geology with Bob Ginsburg. Bob was looking west and describing the vast area of mud on the bank. Meanwhile, Perry kept looking back to the east at the mudflat with algae and mud cracks that lay behind them. Perry had been working on the subsurface geology of an oil reservoir in Montana called the Cabin Creek Anticline. He said, "Bob, the kind of rocks I am working with look more like the sediments behind us." Thus began the study of those tidal flats, according to Perry.

I was still a technician when I made my first trip to Andros Island. Bob had chartered a boat about seventy-five feet in length. Aboard were Ginsburg, Mahlon Ball, and Ken Stockman. In addition, Bob had hired an airboat driver who brought his disassembled airboat. In calm water off the west side of Andros, we assembled the boat and promptly headed for shore. It was one of those flat-bottom aluminum boats with an airplane engine mounted in the back like a big fan, which had been developed for exploration and frog hunting in the Florida Everglades.

Riding the noisy craft at thirty-five miles an hour was a thrill! With it, we could scoot over muddy, intermittently flooded areas inaccessible by boat. It was great fun, and we could all get a "good feel" for the area. It was my first visit of many that would come later. One of the features that stood out was a half-inch-thick layer protruding from the banks of many tidal channels. We saw many of those layers as we whizzed by. None of us knew what they were, but they looked interesting. I collected a few samples but did little with them at the time. We were mainly taking sediment cores that I would later impregnate with plastic in the laboratory back home.

About the same time, another group of Shell geologists, Jerry Lucia, Peter Weyl, and Ken Deffeyes, visited the Pickelmeer—the pink-colored salt ponds on the Dutch-owned island of Bonaire. They had heard that a Dutch geologist had identified the mineral dolomite that was thought to be forming there.[2] Why did they care? I didn't even know what dolomite was, but I learned fast. Dolomite was important to the oil industry because many limestone oil reservoirs were actually dolomite.

On a hunch, I sent a piece of the Andros crust to Ken Deffeyes for X-ray diffraction analysis. I had already put the sample in hydrochloric acid, the standard test for dolomite, but it quickly dissolved away. What I did not know was that the sample was a weak form of dolomite, generally called protodolomite. It is almost as soluble as ordinary limestone.

Ken was excited when he called from Houston. "You have found modern dolomite!" X-ray diffraction had accomplished what acid would not. We had a tiger by the tail, and to help drive things along there was some competition between the Bonaire team and the Coral Gables team. The Bonaire team worked for the production department, and we worked in the exploration department. For years, there had been some tension between the two departments. Exploration decides where to drill, and production does the drilling and testing. If oil was not found, people in the exploration department would say, "The production department screwed up again." It didn't help that the two departments were often on different floors or in separate buildings. Exploration people generally have a more imaginative outlook on things and are often called "arm wavers," whereas production people are more calculating and engineering minded. It is their job to mathematically calculate "reserves" and formation pressures. Fortunately, the tension was not that great between all of us at Shell University. It was about this time that I learned that the laboratory was being called "Shell University" by those out in the oil patch. At that time, one of my closest associates at Shell University was Mike Lloyd. Mike was in charge of isotope geochemistry. If I sent him samples, he would run them. No formal paperwork involved. More than "just tell me the results," he would explain what the results meant. I would be sending Mike samples for many years, and each time I learned something new.

There would be more expeditions to Andros. First, we flew over on chartered amphibious airplanes and landed in the tidal creeks. We chartered Grumman Gooses and smaller planes called Widgeons from Chalks Flying Service in Miami. On one trip we flew over to make a training film for the company. We had with us Ken Hsu, a Chinese-born geochemical genius. Dolomite was one of his specialties. The finished film, as amateurish as it was, became a hit within the company. I would make many films in the future, and I can brag that they gradually became more professional.

On yet another trip to Andros, a summer-hire Texan named Walter Bloxsom and I were dropped off on a dry patch of mudflat with coring and camping gear—and fresh water. We spent five days camping and collecting. Deer flies chewed on us by day and mosquitoes sucked blood at night. During the transition around dusk, no-see-ums, also called sand flies, attacked. They get in your hair, bite your scalp, and slip right through mosquito netting. Fortunately, they sleep at night. For the next trip we made, I assembled a simple plywood catamaran boat that could be taken apart. The two eight-foot-long pontoons could fit in the aisle between passenger seats in the Chalks seaplane. The cat was powered by an outboard motor and allowed us to get around. We named it the R/V *Dolomita*.

During the trip with *Dolomita* we slept in Army-surplus jungle hammocks. Netting kept the mosquitoes away at night, and a canvas roof kept us dry. Best of all, the hammocks were above water when the tide came in. By then, our coring was beginning to reveal the anatomy of the place and showed us how the various tidal-flat zones had formed. Eventually, the company let us go public with our dolomite discovery and I was allowed to be lead author. The presentation was at a national SEPM meeting. As expected, this discovery had a large impact. What made it interesting was that the dolomite was associated with the same sedimentary structures common to many ancient dolomitic rocks. There was no disagreement that this was how much of the ancient dolomites had formed. While we were working on Andros, larger, even more meaningful discoveries of modern dolomite were taking place in the Persian Gulf. Much of what we were learning would pay off during my next adventure—the oil patch and eventually the Persian Gulf. There were new forks in the road ahead.

The Times "They" Are a-Changing

In the early 1960s we heard rumors from the oil patch that more applied research was needed—studies more directly associated with finding oil. As the song says, "The times they are a-changing." Oil-patch geologists were beginning to refer to researchers at Shell University as "eggheads." They thought much of the research was too basic to suit their needs.[3] Some criticism was well founded. I can imagine that plastic shrimp-burrow casts were not much use to them. The Coral Gables office was just a small field office, an offshoot from the main laboratory in Houston. Because we were isolated, we had not heard all the criticism. Besides, all the geologists who visited or took on six-month assignments thought our work had many direct applications.

A Trip to the Oil Patch

After being elevated to the position of geologist, I suggested to Ginsburg that maybe someone needed to go out to the oil patch and see what finding oil was all about. Without hesitation, he replied, "When do you want to go?"

On a rainy winter day in 1963, Pat and I loaded up the VW bug with our three young boys, Boots our black cat, a potted plant, and luggage strapped to the roof, and we headed to Abilene, Texas.

I will never forget bashing through clumps of tumbleweed—I had only seen them in the movies. Some were half the size of my car and rolling across the road driven by a chilling twenty-five-mile-per-hour north wind. No way to avoid them. They literally exploded when hit! The other lasting impression was the color of everything. There was brownness everywhere we looked! We arrived in Abilene and rented an apartment that was half of a duplex. Next day I reported for work. My supervisor would be Dan Bakker.

Dan was the local Shell exploration manager. The first thing he did was put me under the tutelage of my friend Real Turmel. Real had been the first geologist I worked with back in Florida when he was taking cores and mapping the bank around Rodriguez Key off Key Largo. I had already learned a lot from Real, and now he was going to teach me subsurface geology. It gave me new respect for what they do in the oil patch.

Dan Bakker and his wife, Jeanie, took us in as if we were their children. We all had a wonderful learning experience in Abilene. The boys learned about horned toads and goat-head thorns, and they also learned to wear shoes! There was no spearfishing in the bone-dry surroundings, so I bought a mail-order bow and arrow and chased jackrabbits on weekends. After three months in Abilene, we were off to nearby Midland. Midland, which looked like the skyline of Manhattan from a distance, was even drier.

Real Turmel had taught me to describe "cuttings," the rock chips that come up around the drill stem during the drilling of oil wells. Cuttings can be likened to the sawdust one makes when drilling wood or metal. Describing the cuttings on strips of paper is known in the business as sample logging. I made logs of wells in Pennsylvanian and Permian limestone, including those from the famous Horseshoe Reef oil field. At the time, Horseshoe Reef was thought to have once been—a few hundred million years ago—a feature similar to Rodriguez Key bank. So now I was on to Midland to learn about the Permian tidal-flat dolomites and evaporates of the Midland Basin platform.

Again we rented an apartment, and this time I went to work under the tutelage of Duff Kerr. Duff was yet another geologist who had completed an assignment at the Coral Gables Lab. Alan Thompson and a new young employee named Ray Thomasson were two other geologists in the Midland Shell office who influenced my career. With Alan and Ray, I spent weekends helping them map deepwater Ordovician limestone and chert near Marathon, Texas. There was a new kind of deepwater sedimentary structure they were looking for called a Bouma sequence.

One experience I had in Texas made a lifelong impression. It was always clear that our work on how limestone forms might not have direct application to finding oil. Nevertheless, it was thought that there might be indirect application of the ideas we generated. While working with Alan Thompson and Duff Kerr, I learned about a small oil field in the north end of the Delaware Basin. It was called Little Lucky Lake field.

I had given a presentation about results of our fieldwork on the tidal flats of Andros Island and had shown the distribution, and kinds of sedimentary structures, that distinguish tidal-flat environments. We already knew that almost identical sedimentary structures existed in the geologic record and that there had been many giant tidal flats in the geologic past. They are now buried thousands of feet below the surface, and there are plenty under the Midland area—actually the largest in the world! The Permian is by far the biggest oil producer in the Permian Basin, and the San Andres Formation is the largest of the Permian producers. Nearly twenty billion barrels have been produced from the Permian in West Texas, making it a world-class oil producer. The reservoirs are in shoreline sands and buildups, but the seal is what is most important. "Seals" are what keep the oil and gas from leaking upward and out. The seal in this area is "tight"—meaning impermeable—tidal-flat dolomite. Much of the dolomite is similar to the modern dolomite we had found forming on Andros Island. One of the true experts and successful oil finders in these kinds of rocks was Fred Meissner. Fred was one of my inspirations.

Sedimentary structures included features such as algal mats, birdseye structures,[4] storm layers, and mud cracks. The features looked just like those we could see forming on Andros Island. We knew that these structures always formed in the zones flooded daily by the tides or a little higher in what we called supratidal zones that are flooded only by storms. In either case, they were formed very near sea level. In some Cretaceous outcrops in central Texas we even found dinosaur footprints. They were in the same kind of rocks. It was easy to envision the beasts slogging across the mud as I had done many times on Andros Island.

Alan and Duff had seen similar features in cores taken from the Little Lucky Lake field, and they knew that production there was from sands assumed to be beach deposits. If they were beach sands, then the porous body should trend parallel to the ocean (in this case the ancient Midland Basin, which was an inland sea during Permian time). That's the same layout one sees on any beach today. The problem in this case was that most of the geology in this oil field was based on electric logs, especially a "kick" on the logs dubbed the "X-marker." The X-marker was about five hundred feet above the oil-producing zone, so it did not provide the accuracy needed. What was needed was a marker indicating ancient sea level right within the oil field.

Alan and Duff remembered they had seen sedimentary structures in some cores that looked exactly like what we had described on Andros Island. Knowing that tidal flats indicate sea level, they took another look at the cores and compared them with the electric logs. There is no better horizontal time-line marker than sea level. Sure enough, there was another electric-log marker coincident with a horizon containing mud cracks and laminations made by tidal-flat algae. Here was an ancient sea-level marker almost as accurate as a spirit level used by carpenters.

The best part was that this marker was right within the oil-producing zone. With this in mind, Alan and Duff quickly recognized that the sandy producing zones were not preserved beaches. The porous oil-saturated sands bowed downward, like sands in a channel. They were more like what we had been finding in modern tidal channels on Andros. We had also shown that the tidal channels on Andros contained sandy sediment, but, most important, the channels are aligned more or less perpendicular to the shore, and they were often interconnected. All of this changed their thinking about where to drill the next production well. And they struck pay dirt!

Being tidal channels meant the targeted structures should be *perpendicular* to the trend envisioned when they were thought to be beaches. That's a 180-degree turnaround! The new concept had been sparked by our work on the tidal flats of Andros! Here was uniformitarianism at its best! Now the reader can appreciate why this made such a lasting impression on me. Considering the cost of drilling oil wells and the money gained from the production, the salaries of us "basic egghead scientists" seem trivial.

That oil-field experience also made a lasting impression on me about how people think or don't think. People can become fixed on a single idea and somehow fail to see alternatives. Sometimes it only takes a chance occurrence, a tiny spark, to completely turn things around. I have seen it many times. I liken it to the homily, "When you are on the wrong bus, every stop is the wrong stop."

Because there were also dolomite and evaporites associated with the oil field, we decided an even better comparison could be made with the tidal flats (called sabkhas) on the southern shores of the Persian Gulf. I didn't know it then, but soon I would be in the Persian Gulf slogging across the world's most famous tidal flats and sabkhas.

Weekend Adventures

A Florida family living in Abilene and Midland, Texas, can have a lot of fun even if there is no water. In a few hours we could drive to Carlsbad, New Mexico, and the world-famous Capitan Reef, an ancient reef now exposed in the mountains made famous by geologists Lloyd Pray and Philip King. And there were Carlsbad Caverns and White Sands, New Mexico, where the first atomic bomb was tested. We could also drive to Big Bend National Park, or we could cross the border into Mexico where tequila and boots were cheap. At other times we would drive south to the Hill Country in central Texas and look for dinosaur footprints that are exposed in rocks in several creek beds. Our boys loved that!

I spent several weekends crawling over the Paleozoic algal mounds in the Sacramento Mountains near Alamogordo. Those mounds, sometimes called "algal reefs," are exposed on the mountain slopes and are well known because of research by James Lee Wilson, a famous researcher and teacher at Shell University. Many years later, I would return there with a portable

West Texas days: Patricia, Gene Jr., Tom, and Dennis with Permian El Capitan Reef peak in background on one of many weekend excursions enjoyed while Gene was stationed in Midland, Texas, in 1963. Permian rocks date back to between 299 and 251 million years ago. Photo by author.

Dry-land substitute for spearfishing. Gene pursued jackrabbits and Texas-size frogs near Abilene, Texas.

coring device to test some ideas. These weekend trips prepared me for exciting adventures to come.

On short weekends we would simply go play with our three young boys on the sand dunes at nearby Monahans outside of Odessa or collect cactus with Ray Thomasson for his garden. According to Ray, there were two kinds, "two-man" and "one-man" cacti. It took two men to get the former back to the car. I could also go out with my new bow and arrows and chase Texas-size bullfrogs that congregated in ponds along dry creek beds. The frog legs were delicious—just like chicken of course!

A short drive out of town would take us to caliche pits. Caliche is a kind of cap rock that forms under desert conditions the world over. It is a hard cap rock mined for use in road construction and often contains marble-size concretions and layers that are often bowed upward to make features called *tepee* structures. Knowing something about caliche would become important when we went to the big deserts in the Middle East. Weekends in West Texas were like going to the beach, except there was no water. The sand dunes and all I learned there would prepare us for our next big adventure. We would really experience desert beaches, but this time there would be lots of very salty water.

3

The Netherlands and the Persian Gulf

A Big Move

By the mid-1960s, Royal Dutch Shell had recovered from World War II and had established a research lab at Rijswijk, a suburb of The Hague, near company headquarters. Geologists at the Rijswijk Laboratory, or Kon. Shell Exploratie en Produktie Laboratorium (KSEPL), had been intrigued by our research on modern carbonates in Florida and the Bahamas. A decision had been made to do similar research in the Persian, now "Arabian," Gulf. The geometry and climate of the Gulf Basin resembled those of many ancient carbonate basins, especially the Permian Basin in West Texas, which were prolific oil and gas producers.

Whereas the Bahama Bank is an enormous carbonate bank with high relief surrounded by deep water, the Persian Gulf is truly a basin. It consists of an axial trough, flanked by coral reefs, tidal flats, salt domes, extensive areas where dolomite is forming, and evaporates, including anhydrite. Anhydrite, normally found at depth, had never been found at the surface, but Douglas Shearman had recently observed it forming on the sabkhas near Abu Dhabi. The Persian Gulf tidal flats also have extensive algal mats and quartz-sand dunes. Here was everything a sedimentologist could ask for.

After preliminary work by Dutch sedimentologist J.J.H.C. Houbolt, Bruce Purser was hired to lead the extensive project. Bruce, a robust New Zealander, was raised on a sheep farm, played rugby, and had worked in primitive jungle conditions in Southeast Asia. Surprisingly, his true love was butterflies and all insects. He was not only a geologist but also a true naturalist and would later publish books on butterflies.

Bruce had been put in charge of Shell's large Persian Gulf program that would be organized from KSEPL in Rijswijk. He recruited many fine geologists, such as Brian Evamy, Mike Hughes Clark, and Peter Kassler from England. Gerry Varney, also a Brit, was brought in from Shell Canada, and there was a young Dutch geochemist named Kees DeGroot. Bruce also worked closely with the Imperial College group, Graham Evans and Douglas Shearman, and there was Professor E. Seibold from Germany. Douglas Shearman already had a group of students working in Abu Dhabi, one of whom I would collaborate with many times in the future.

Bruce had already made preliminary trips to the area with Ken Glennie, Bob Ginsburg, and James Lee Wilson. With all this geological power behind him, KSEPL, under the leadership of Ludwig Happel, made the decision to launch an expensive multiyear expedition to the Gulf. This would be the new "Shell University." Because I could dive, run boats, core sediment, and fix things when they broke, I was asked to join in. My father, who was then living in the Philippines, had encouraged me many times to go abroad if the opportunity was presented. He had taken many forks in life's road, and many more lay ahead for him. The Shinns were ready!

With Patricia and our three young boys, Gene, Tom, and Dennis, in tow, we arrived by plane in The Hague during the winter of 1965. It was cold and raining, a typical day in Holland. The climate was very unlike West Texas, but at least everything was green, much like Florida but without palm trees. After settling the family in a flat in Rijswijk, I set off for my first expedition to the Gulf, reluctantly leaving my family behind in Holland.

There were mountains of equipment to be sent, and no expense had been spared. To gain an overview of the area, four of us—Bruce Purser, Ken Glennie, Mike Hughes Clark, and I—made a complete tour around the Qatar Peninsula shoreline. We were looking for promising study sites with varying sedimentary environments. We were also recording the aftermath of a severe storm, a hundred-year shemal that had killed fish, dugong, sea snakes, and coral reefs a few months earlier in 1964.

Alan Wells and Leslie Illing had already located many interesting areas where dolomite was forming, and there were extensive algal mats to be investigated as well. We gained a good understanding of the area by camping as we toured with our British Army–style Land Rover towing a trailer with a Boston Whaler outboard boat. We camped and made several brief trips offshore to look at mostly dead reefs. With us was an interesting Indian cook who, years earlier, had worked for the British military back in India. He told me that one of his duties had been to shave the beards of British officers before they got out of bed in the morning. I had purchased a battery-operated electric shaver and decided to shave myself.

When not in the field, we stayed in a flat in Doha where we stored equipment for the larger Gulf-wide cruises that were to follow. I thoroughly enjoyed the area—living conditions in Doha were much more to my liking than the rain, snow, fog, crowds, and rigid social restrictions back in The Hague—but being away from my family for three long months was more than I could stand. When our team returned to the Netherlands, I made a decision to return to the States unless Shell would allow me to move my family to Doha. We were not conditioned to spending that much time apart, although I had met many Europeans who had spent years away from their families. Management agreed, and we all moved to Doha. I think the move also saved the company money, and besides, I could keep my eye on valuable equipment that was stored there.

By the time we moved we had spent six months in Holland, where the boys attended International School. The oldest had broken his leg, but it had healed. The next two years would be the adventure of our lives, professionally, socially, and financially. We would pay no American taxes, no Dutch taxes, and we could afford a house with a servant! I had the use of all the research equipment, including two Land Rovers, rubber boats, a Boston Whaler, tents and camping gear, diving equipment, scuba tanks with an air compressor, and microscopes. I even set up a photographic darkroom and a lab for impregnating sediment cores with plastic resin just as I had done back in Coral Gables.

Because the KSEPL research operation was a separate entity from the local Royal Dutch Shell Operating Company, they could only provide me with some tactical support. I was to be a one-man operation. The local Shell Company, however, had established a system—the Bank of the Middle East—for funneling KSEPL resources, and my salary, to our operation. The operating company also had some fine petroleum geologists who had an interest in what we were doing. One who accompanied us on many field and weekend camping trips was Australian geologist Bob Foster and his wife, Jan. Bob was a natural-born collector with a great sense of curiosity. We had met back at the Coral Gables Lab some years earlier, and he had helped me with my coral-growth experiments.

The Shinns were the only Americans living in Qatar at that time, and I was professionally pretty much on my own. Some Americans, usually from geophysical companies, would visit briefly and then be on their way. The non-Arab community of about five hundred consisted mainly of Dutch, German, French, and English citizens. There was, however, an abundance of Pakistani, Indian, Baluchistani, and Palestinian manual workers. They outnumbered the Qatari Nationals, and it remains that way today. In fact, at this writing, Qatari Nationals are totally outnumbered by many nationalities.

At that time, Qatar was a British Protectorate along with the Trucial Coast, now the United Arab Emirates. Qatar is separate from the Emirates but is closely allied with their Emirates neighbors. Both Qatar and the Emirates have a testy relationship with their biggest neighbor to the south—Saudi Arabia.

Most of the Shell Operations people lived in a compound. In the beginning we lived in the same building where our equipment and office were located. An adjacent apartment or flat served as our laboratory/office and storage area for equipment packed in huge wooden boxes. The heavy equipment was stored within the walled parking compound behind the apartments. Taking care of the equipment was part of my responsibility. The heavy equipment—a heavy bomb-shaped "Houbolt corer," a box corer, as well as an electric vibracorer—would be loaded onto a large chartered vessel once a year when the rest of the team came down from Holland to conduct Gulf-wide seismic profiling and collect sediment samples.

High-resolution seismic profiling was achieved with a sparker system, and navigation was accomplished with Decca navigation that had been installed in the Gulf especially for offshore oil exploration and production. The Global Positioning System (GPS) was years in the future. I participated in the Gulf-wide cruises for one or two months each year, but for the remainder of the year I was free to map and core lagoons and tidal flats around the Qatar Peninsula. Professionally, it was the highlight of my career. I was thirty-one years old and about to make a discovery of a lifetime. I had also hired a helper who would become a lifelong friend.

Social Life in Qatar

The Shell Operating Company maintained an English school consisting of four contract teachers from England. Our two youngest boys, Tom and Dennis, attended the Shell School. Our oldest, Gene, was nine, the age that most English parents send their children back to England for boarding school. Because he could not attend the Shell School, Pat homeschooled our nine-year-old fourth-grader with a teaching kit purchased from the Calvert school system in Maryland. It turned out to be an education for all of us. Gene would finish his work by noon and then had the rest of the day off. There was plenty of desert land for him to explore with his free time. If I wasn't in the field with my helper or on one of the Gulf-wide cruises, I was in the lab preparing and examining samples or writing. I had two Land Rovers, and for Pat we purchased a used Austin mini, a little car that was very popular in England.

On most afternoons we went to the Doha Sailing Club, where there were sailboat races twice a week. It was interesting to see how the British and Europeans entertained themselves in faraway outposts such as this. I always liked to say, "It's like something out of the movies." We fell right into it.

In addition to sailing, there was a skeet club, where you could shoot clay pigeons, and golf that was played on rocks and sand. The "greens" (they called them browns) consisted of sand kept smooth with a coat of crude oil. I didn't play golf. On the other side of the peninsula at the International Petroleum Camp (IPC) there was yet another sailing club, and the Brits over there had built a go-cart track. On occasions we had sailing competitions with the IPC sailing club.

When we joined the Doha Sailing Club, we were not allowed to use the boats until I passed a sailing test to obtain my little green sailing card. It was all very proper. Fortunately for me, the commodore was a Dutch geologist at the Shell Operating Company.

I had grown up on and under the water, but it had always been with motorboats. I had never actually sailed. Now I had to pass a sailing test. The German fellow who was in charge of the boats took me out for my test. I thought I did very well. Sailing is pretty straightforward if you are comfortable on the water and know the feel of small boats. Nevertheless, the examiner failed me because I did not tie the anchor line with a bowline knot! I appealed to the commodore, who personally conducted the next test. (In the meantime I had learned to tie a proper bowline.) The commodore of course passed me, and I got my sailing card. Two days later, Pat and I competed in our very first sailing race. We all sailed on the same type of boat, a fourteen-foot two-person sailboat called a Kestrel. The big highlight was we beat the German fellow who had failed me! It was a storybook win, and the word quickly got around. From then on, we were fully accepted.

Socializing was intense because remoteness tends to draw people together. We were all from different parts of the world but had a common bond, oil and survival. The Operating Company maintained a "lodge" for its employees, and as fellow Shell people we were always welcome. The lodge provided cinema twice weekly (the films were not that good, but it was a diversion). The films that drew the most laughs were the American Hollywood epics centered on Texas! It was most interesting watching the audience responses to the films.

There were darts and billiards, and on any evening we could enjoy a delicious meal presented by Indian servers wearing white cotton gloves. We could also purchase liquor very cheaply. Because it was a Muslim country, only Europeans were allowed to consume, and if you were a real party person you could participate almost every night of the week. There was plenty of booze. Interestingly, we never encountered anyone whom I would consider drunk. Maybe I was and did not notice; I do remember morning-after hangovers.

After a few months we had heard all the best stories people had to tell— and at least a dozen times! The social highlight was the arrival of any new foreigner, which would be an excuse for several more parties and would mean a round of new stories that would last us for about a month. "Fresh blood" I called it. There were some advantages of being the only American in this community.

One night we were invited to a unique party arranged by the British Embassy, which was being held for the CEO of U.S. Steel. He quickly gravitated to us, the only Americans. "What's it like for you as Americans to be living in this faraway place?" he asked. He was very interested in anything we had to say. I'm sure he felt just as out of place as we, so we were naturally drawn together. A meeting like that wouldn't happen often anywhere else.

There was yet another party, a strange one in fact. We went to a gathering of Shell Operating Company people to meet a visitor. I had no idea what he did, and I didn't ask. The shock came a few days later when I learned he was a diviner, or what many Americans call a "water witcher"! He was en route to Oman, where Shell was building a new camp that would need freshwater. I couldn't believe it. Here we were, a supertechnical company using the most sophisticated equipment in the world to find oil, and a diviner had been hired to find water! I learned later that the fellow did find water, but it was salty. I was glad I didn't ask about his profession at the party.

A useful lesson we learned early on was that it is impolite to discuss business in a European or British social gathering. Business discussion at parties is a typically American thing. We learned that lesson well and still avoid business discussions at social gatherings when possible.

Camping was a favorite weekend getaway. Camping also allowed additional observations of the land and seascape. We would take the Land Rover or the mini and drive in any direction, except due south—that would take us to the Saudi border—and emerge somewhere at the water's edge. We had many favorite places, and there would often be several couples. Fish were abundant. I could start the fire, swim offshore, and spear a five- or ten-pound hamour (grouper) within ten minutes and have it in the frying pan twenty minutes later. It was wonderful! In fact, the spearfishing would lead me to a paradigm-changing discovery.

Shopping for food was always an adventure. There were only two so-called cold stores in the city. A cold store was where you could purchase frozen meat, butter, canned goods, and some vegetables. Most was shipped in from Australia or New Zealand, and availability was determined by how recently a cargo ship had landed. It was the same for other items ranging from screwdrivers to hammers and any other needed hardware. Availability of anything at the *souk* (local market) also depended on when the most recent ship had arrived. When new necessities hit the market, word would spread through the European community like wildfire.

Most every young Qatar National smoked cigarettes. A typical scene was a young teenager smoking a cigarette with one side of his headdress pulled

A typical camping trip evening in the desert, 1966. Pat and Gene with drum and British friends Russell and Theresia Grosch.

Shinn boys Gene Jr., Thomas, and Dennis on a camping trip in western Qatar.

to one side. I guess they thought it was cool. Several brands of English cigarettes were sold at the cold store, but they were all called Ali Binallie cigarettes. That was simply because they were all purchased at Ali Binallie's Cold Store.

A favored pastime was shopping in the souk for handmade Persian carpets. There seemed to be an endless supply and a multitude of varieties. Carpet shopping was an adventure even when not buying, and we did a lot of that. Pat had her eye on a particular Shiraz carpet and would haggle with the merchant every time we visited the souk. One day the owner was not there, so she haggled with the owner's son and got the price down to what she wanted, paid the boy, and left as quickly as she could. She was hardly back to the car when the owner, who was furious, came running after her—she escaped just in time. That carpet still lies beneath our dining table.

The 1960s was a decade when Arab nations were attempting to emulate the West. Many Arab men wore Western clothes, and automobiles were rapidly replacing camels and donkeys. Nevertheless, many old ways remained. Most houses were made of mud and stone. Some even had outhouses on the second floor overhanging alleys. In the dry climate, noxious odors were rarely noticeable because everything dried up and blew away so quickly. Not so for the flies—they were everywhere and were constantly landing on your face or crawling over your sunglasses. I could never find where they were breeding!

The capital city of Doha, especially around the main souk, was a constant cacophony of jingling bells, baying goats, and car horns, especially Italian horns that sounded like trumpets playing bugle calls. Horns sounded at every intersection, often for no apparent reason. Honking seemed to be a constant nervous habit. On top of all this, Middle Eastern music blared from loudspeakers everywhere, "just like in the movies"!

Mixed in with all this cacophony of sounds and veiled Arab women were European women wearing miniskirts! Even Pat wore them. At the same time, Muslim women were completely covered in black as many remain today. It was an eclectic blend of mixed cultures. However, with the return to fundamentalism in the 1970s and 1980s that all changed. Today, most Arab men take pride in wearing traditional clothing, which, after all, is more comfortable and more suited to the climate. Women still wear the abayas but can now show their faces. Older traditional women continue to wear the mask or at least a black veil. There are no miniskirts in public today. However, Western women around the pools of the many exclusive hotels wear bikinis. Hotels in the area today are unbelievably modern by Western standards and well kept.

About a year into our assignment, we left the children with a friend and took a side trip to Iran. There we rented a car with driver and toured several small villages to watch the weaving of handmade carpets. By then we had fallen in love with Oriental carpets. What we saw in the villages revealed that mainly small children really were doing weaving, as we had heard. There were many famous mosques to visit, but the highlight was the drive to Persepolis, the ancient cultural center of Iran. Persepolis had been constructed around 500 B.C.

Another side trip was to Cairo, and of course the great pyramids and the traditional photos of us seated on camels. We hired a grandfatherly guide who completely captivated our boys. He took them in the tunnel that leads to the burial chamber deep within the Great Pyramid. Next he guided us through the Cairo Museum—we got our fill of stories and mummies. Everyone we met was friendly, and we felt safe everywhere we went. We did discover when we left that their money was worthless outside of Egypt.

The next stop was Beirut, Lebanon, known at that time as the Paris of the Middle East. With yet another native guide, we were driven over four-thousand-foot snowcapped hills to the remains of Baalbek, the great Roman city built over earlier ruins of Heliopolis dating to the time of Alexander the Great. What mystified me most as a geologist were some giant pink granite columns. Where did they come from? They had been imported from Aswan, Egypt, the closest source of that type of granite. Somehow they had been transported down the Nile and eventually across rugged high hills. I have great respect for the accomplishments of the ancients and their primitive tools. How could they remain so motivated to build such monuments for a period of hundreds and even thousands of years?

From Beirut we moved on to Constantinople, Turkey, bordering the Bosphorus, the waterway connecting the Mediterranean with the Black Sea. While there I went outside the Hilton Hotel and changed American Express Travelers Checks for Turkish money with the ever-present money changers. The illegal exchange rate was so good that I profited enough to pay our hotel bill!

Here was yet another giant bazaar and abundant brass objects to barter over and where we were pestered constantly by the ever-present shoeshine artists with highly decorated equipment boxes. Giving in was the only way to get rid of them—until you rounded the next corner. Wearing sneakers kept them away! It was an absolute must that we tour the Blue Mosque and the Topkapi museum, the former Ottoman palace with its many harem rooms.

These trips, along with a visit to Rome, the Coliseum, and the Vatican, had been lifelong dreams. I can say with pride that at no time during these cultural side trips were we part of an organized tour. On many occasions we found it embarrassing to be near and overhear the conversations of typical American tour groups. The German tour groups we could not understand, but they were everywhere taking photos along with counting and measuring everything with tape measures for their notebooks. We made a concerted effort, and for the most part successfully, to fit in to the culture everywhere we went. The only thing that gave me away was my ever-present camera, which I learned to keep hidden in a paper bag. You become a target for every con artist in the world once you sling a camera over your shoulder.

What stood out then, and still does today, was the friendliness of the Arab people. In all the Arab cities we visited, especially the villages, there was little crime, and to my knowledge there is little today. It was a different story, however, if you left a car in the desert. Tires and battery would be missing when morning came. If left for a second night, most everything else would be stripped. I had that experience once when I ran out of gas. Luckily, I was back early the next day—only the battery and a spare tire were missing. Justice was, and remains, swift and effective, and in addition, most homes are enclosed within high walls often with broken glass decorating the top. Swift justice means you can lose a hand, and I occasionally saw men with a missing hand. Most were reduced to begging.

Men and women still lead segregated lives in public. They eat separately in restaurants and attend social events in segregated areas. Dentist offices, for example, have separate rooms and chairs for men and women. In the 1960s, marriages were mostly prearranged. That has changed somewhat. Today, urban surroundings are ultramodern. Ice skating and snow skiing in the mall—need I say more!

A Controversial Discovery

In Qatar I would make an unexpected but important discovery, one that would alter my career. It also taught me a lot about the reception, or rejection, of new scientific discoveries and ideas.

On the west side of Qatar Peninsula there is an old jetty near the village of Zekrit. We camped there one cool autumn weekend, near the jetty, where we could feel the warmth of a giant gas flare at night. The flares were near the north end of the Ducan oil field. Natural gas was essentially worthless then, so it was flared. There were dozens of truly giant flares that reached one hundred feet into the air. They could be seen at night from the other side of the peninsula.

In the afternoon near dinnertime, I headed for the old jetty, put on my dive mask, and eased into the seven-foot depths at the far end. In almost no time I had speared dinner, but the fish swam with my spear under a rock ledge. While removing the fish and spear, I noticed pieces of pottery, glass, nuts and bolts, and other artifacts that had become part of the rock. These objects had been jettisoned from dhows that had once moored there. Out of curiosity, I broke off a piece that contained pottery. I was immediately excited! Why? The ruling paradigm of the day, especially within the Shell research community, was that lime sediment could only become limestone when exposed to freshwater. These artifacts clearly indicated that this rock was young. I knew the jetty itself was only thirty years old, and there was no way this rock could have been exposed to freshwater!

In my little lab back in Doha, I made a petrographic thin section of the rock with the pottery and examined it under the microscope.[1] The fringe of aragonite crystals that, in essence, bonded individual sand grains together also surrounded the pottery! The pottery truly had become part of the rock. I had seen other ledges of rock in the Persian Gulf and assumed, as we all did, that it was rock that had formed under freshwater conditions when sea level was lower and the rock was exposed to the atmosphere. We believed that it had to have happened several thousand years earlier when sea level was lower. Let me explain.

One of the great disappointments we faced during our first Gulf-wide survey in the large chartered boat was the scarcity of sediment. We had to drop the sediment grab several times to bring up a small sample. It was worse when we cored. The coring device consisted of a heavy, bomb-shaped weight with a steel tube attached to the end. It would bounce off the bottom and come up looking like an accordion! We had begun to believe that there had been very little sedimentation since sea level rose a few thousand years earlier. What were we going to do? This was an expensive venture. After all, we had come all this way to collect sediment! The whole Persian Gulf project was based on studying the distribution of sediment, and we were not finding much to study. My accidental discovery after spearing a fish implied a simple explanation.

The spearfishing incident and the rock with young artifacts indicated that the rock below the jetty had formed underwater. Would this really be the explanation for all that rock out in the Gulf? Of course, this could not be! Or could it?

Excited, I wrote up my observations and sent the note and photos back to Bruce at the KSEPL in Holland. The documents received skeptical reviews. "Possibly cement used for jetty construction had created the rock," Bruce and others said. Or, "Possibly some chemical had been jettisoned from a moored vessel along with pottery and glass bottles."

I went back to Zekrit for another look. Several kilometers away, where no jetty or harbor existed, I made another discovery—more pottery cemented into the rock! This time it was quartz sand that had been converted to sandstone. The bottom was ornamented with sand ripples. Ripples mean moving sand and water currents. However, these ripples were stationary and hard, as if they had been frozen in place. Right then, I knew I was onto something big! It could explain the near absence of loose sediment in the Gulf. Fortunately, I was totally free to go where interest and instincts led me. My boss was several thousand miles away and five days by mail.

Across the lagoon from the Zekrit jetty there was an attractive campsite with a small, protected sandy beach. From a previous camping trip, I knew the bottom was smooth and hard just off the sandy beach. With a forty-pound chisel-shaped iron bar purchased in the souk, I banged away at the hard bottom. With some effort I broke through a layer of rock and came to sand underneath. The rock was a four-inch-thick layer of cemented sand. Here and there the layer was coated with freshly precipitated amber-colored aragonite. Aragonite is the mineral that was binding the sediment grains together, and it forms only in seawater or in saline water in caves.

Aragonite is the same mineral that corals and many marine shells are made of. Clearly, this aragonite, this special form of calcium carbonate, had precipitated directly from the seawater!

Where there was no sand blanketing the rock, the rock was exposed directly to water. In those places, the rock was riddled with borings and encrusted with calcareous worm tubes. A little more digging with the heavy wrecking bar revealed another rock layer below the first—and then another! After more digging, I had uncovered four separate rock layers, each separated by loose sediment. Because submarine-rock creation was not supposed to happen, I spent several days at this place digging rows of pits leading out to deeper water. The pits allowed me to trace each individual layer. Farther offshore, in about ten feet of water, the rock layers ended and merged with soft, almost jelly-like sediment. It was clear the sandy sediment was being converted to rock while the muddy fine-grained sediment was not.

I submitted a sample for carbon-14 analysis and waited expectantly. My heart was pounding with excitement when the date proved the rock too young to ever have been exposed to freshwater. Besides, freshwater is really scarce in this part of the world. Now I knew I was onto something revolutionary!

Next I sent some samples to Mike Lloyd back at the Houston laboratory. Mike had moved to the main Houston lab after finishing his PhD and had established his own isotope lab. He too was curious and quickly ran some isotope analyses. His oxygen and carbon isotope results ruled out any exposure to freshwater.

Meanwhile, letters from headquarters back in Holland reminded me that I should be working on sabkhas and tidal flats, not chasing layers of rock in the Gulf. Nevertheless, the possibility that there really was rock forming in the Gulf began to catch hold at the Royal Dutch Shell Laboratory. Regrettably, that was not happening back at the Houston laboratory. I got back to my sabkha work but combined weekend camping trips with the family with chasing rock layers around the Qatar Peninsula. It was reminiscent of weekends in West Texas. The main difference now was saltwater and fresh fish.

It wasn't long before I found that submarine-cemented limestone was also intimately associated with the sabkhas I was studying. Layers of this

Core drilling the sabkha with a Royal Dutch coring device at Umm Said in southeastern Qatar.

rock were preventing vertical fluid flow within the sabkha environment. They also interfered with sediment coring! Could this knowledge aid production geologists exploiting ancient tidal-flat deposits? It soon became clear that my discovery of submarine rock was not a popular idea back at Shell University in Houston. Bob Dunham had already convinced most of our oil-patch geologists that freshwater was necessary for conversion of sediment to rock. The seeds for a long and often heated controversy within the company were being sown.

With the controversy in full bloom, it became ever more important to thoroughly investigate and prove to the company geologists that, without a doubt, marine cementation was happening and should therefore be expected in ancient limestone. Uniformitarianism again! What were the criteria for its identification? If we were going to identify this kind of rock in ancient rocks, we needed to know exactly what to look for.

One of my more critical finds came while diving with Arab pearl divers. To the east of Qatar lay the Great Pearl Bank—a place made famous by pearl divers and the pearls they had marketed for centuries. This was the historical site of the legendary Persian Gulf pearl trade. This bank is a large area of nearly flat bottom consisting of almost bare rock. It covers hundreds of square miles. Depth ranges from thirty to one hundred feet. It's where we had bent so many core tubes trying to take sediment samples.

The pearl oyster, not really an oyster, requires hard rock for attachment. It does not live in or on mud or sand. While diving with the pearl divers, something I did on many occasions, we would come across fissures in the rock. The fissures were actually the edges of huge polygon-shaped layers of rock. The water depth was about sixty feet, and the rock was thick and extremely hard. The question was, how could I collect a large piece of this hard rock? It was too hard and too thick to sample with the iron bar. Flashback to a former occupation: an "explosure" was needed!

Opportunity knocked when an American geophysicist from a seismic-survey company happened to be passing through Doha. A party was thrown for his group. After a few drinks, we developed a plan. Suppose a few seismic boosters somehow got lost on a beach somewhere? And suppose I should stumble across them. What a find that would be! Yes, I did find them and they were seismic boosters, the same kind of small chargers I had placed in the nitroamone explosives back when I did the summer job in Cuba.

So with lamp cord, a battery, and my assistant, we ran the Boston Whaler offshore to about sixty feet of water and found a suitable fissure. Placing the charge in the fissure was reminiscent of my time on the salvage boat. At that depth the small charge made only a muted thump, but what it did was spectacular! It revealed a layer of rock a foot thick that was so hard that every geologist who saw it said, "It looks like typical 100-million-year-old limestone." However, carbon dating showed it was only three thousand years old! It was much too young to have ever been exposed to freshwater.

The rock also contained pearl-oyster shells. The rapid submarine rock-making process had created the substrate needed by pearl oysters. Now that was really interesting! It also meant that the rock-making process could explain certain fossil distributions during the geologic past. In other words, the substrate controls the distribution of organisms that make the sediment

Cross section of a one-foot-thick layer of newly formed rock exposed by the use of an explosive seismic booster in sixty feet of water on the Great Pearl Bank in the Arabian Gulf. Object in foreground is a geological rock hammer.

that makes the needed rock substrate. No one had considered that the rock-forming process itself could also provide the substrate needed by fossil organisms. This was a new way of thinking about carbonate geology. A new example of uniformitarianism!

There were many other implications. Bored surfaces (holes made in the rock by rock-boring organisms) had generally been interpreted as evidence of exposure. Many bored surfaces are evidence of exposure. The trick is to tell them apart, because exposure would thus indicate past sea-level fluctuations. Our rigid paradigm-driven interpretation at that time held that "bored surfaces must indicate that there had been a drop in sea level." That a rock could be forming that mimicked changes in sea level flew directly in the face of a new emerging paradigm of geology called "sequence stratigraphy." Not that sequence stratigraphy is invalid; it's just that the geologist could get the interpretation of a rock formation backward if all bored surfaces were interpreted as evidence of a drop in sea level.

And there were yet other implications. There was no dissolution involved in the formation of this submarine rock. Without dissolution, this rock would have very little porosity. A rock formed this way might never obtain or preserve enough porosity to be a useful oil reservoir, but it could act as a seal and thus help to trap oil. It was a bittersweet double-edged discovery. We were supposed to find processes that make reservoirs, not destroy them. At the same time, we needed this knowledge in order to avoid those rocks that can't hold oil. Knowing where not to look might help lead to finding the sweet spots—those places where there is oil.

The next step needed was to determine how extensive this process had been throughout geologic time. Such knowledge could possibly help others find, or not find, oil. It all depended on the chemistry of ancient seas and many other factors. We needed a larger view.

My Trusted Assistant

Early during my stay in Qatar, I had a flat tire in the desert. I was jacking up the Land Rover when a car pulled up and, with a heavy accent, the Arab fellow said in broken English, "You need help?" It was the beginning of a long friendship. His name was Rashid Al Rachman. He was a native from the village of Khor. Rashid owned a taxi. He had worked for the Shell Company and knew just enough English to communicate. He was very friendly, and, as I would soon learn, very trustworthy. The next day, Rashid showed up at my little office and offered his services. The pay wasn't much, but he was willing. Through him, the whole village of Khor soon knew me, especially the children, and became my friends. If I wanted something, Rashid knew where to find it. I could draw a picture of a screw, and he would ferret them out at the souk. Same goes for screwdrivers, hammers, you name it. For example, Rashid had found the heavy bar I used to sample rock layers underwater. If I needed to dig a trench in the sabkha, he would find a team of diggers. He also knew all the dhow owners.

Rashid was about my age. His exact age is still not known for sure. He had a five-year-old son and at this writing Rashid is well into his seventies, and his son, Mohammad, is in his late forties (now Dr. Mohammad with a degree in physics). We stay in touch over the Internet and see each other once every few years. With Rashid's help, we could find most anything, go anywhere, and arrange anything from a camping trip, to a meal, to chartering a native boat.

Dhows for Two

During the final year of research, Bruce and I decided that instead of having everyone on a single large Western-style steel vessel, we could split up into two teams using native dhows. Dhows were much cheaper, and we could go in shallow water not accessible to large ships. There was another advantage that became apparent once we were at sea.

With Rashid's help we soon arranged to charter two diesel-powered dhows, each with a crew of about ten. Bruce would take one dhow, and Gerry Varney and I would take the other. We would split up and cover more territory at less cost. This would be one of my most exciting adventures in the Persian Gulf.

Gerry and I would do systematic bottom sampling using a Decca navigation system for location. The Decca system, set up for the Persian Gulf petroleum industry, was similar to modern GPS. Instead of satellites, it relied on signals from strategically located radio towers. By plan, we would work mainly west of the Qatar Peninsula in the Gulf of Salwa and around Bahrain, while Bruce would work east of Qatar off the Trucial Coast.

One of the unique advantages of working from a dhow became apparent once we came near a foreign shoreline. Dhows can go anywhere! There are no national borders for them. When we neared the Saudi coast, the captain would haul out a Saudi flag. A European vessel could not do that, and it would take months to negotiate formal entry into foreign waters with any other type of research vessel.

On board we had my diving equipment, including a gasoline-powered scuba compressor. That was easy. However, the Decca navigation system required electricity and a cool working environment. For that we installed a diesel generator in the forward hold. A jerry-rigged length of core tubing over the exhaust carried sound and fumes twenty feet above the deck. We had electricity! Next we needed a cool environment. Using plywood, we constructed a ten-by-ten-foot white box on the aft deck and installed an air-conditioner to keep it cool.

Our little house on deck was for the electronics, but we quickly learned it would be a good place to sleep during hundred-degree high-humidity nights—it was hot every night. The crew slept on deck beneath the stars as dhow crews had done for centuries. They followed a curious rigid ritual.

By day we all worked beneath a heavy cotton sunshade that covered most of the boat. We called it dhow cloth. It wasn't waterproof, but that didn't matter because it never rained. However, at night the crew took it all down so the dew would settle on the decks—including those who slept there. We never quite understood the ritual and never could get a direct answer. It just seemed like something that had been done for centuries and that's the way it was going to be. I suspect they liked looking into the heavens. Maybe it was also to keep the teak decks from drying out and cracking.

While Bruce investigated beachrock and reef areas to the east, we went west and made many interesting observations. There were strange polygonal-shaped ridges on the seafloor near Bahrain. The ridges were made by the expansion of thin rock layers. They were the same kind of layers I had found off the beach in Qatar. The process of converting sediment to rock caused expansion, and expansion had pushed up ridges much the same way ice in the frozen north buckles to form ridges. The features looked like the so-called tepee structures in the caliche I had seen back in Midland, Texas. They were also like those in the Permian rocks of Texas and New Mexico. The Permian features had long been interpreted to be like those in caliche. Think of an old cement road or sidewalk buckled up from expansion caused by heat from the sun. However, what we found was forming below water! That was different. Fortunately, I had looked at many tepee structures during my stay in Midland. I might not have realized the significance of these underwater ridges had I not seen them in Texas. A few years later, we would find that many of the same features in the Permian rocks in New Mexico had been formed underwater like those we were finding near Bahrain. Today, they are well known and can be distinguished from those that form from caliche on land.

One of the favorite meals aboard the dhow consisted of small sharks the crew caught at night. The crew put them on the foredeck and left them in the sun by day. The sun treatment baked out much of the urea smell and taste that usually identifies shark meat. In the evening, pieces of shark meat were boiled in yellow saffron-stained rice and served on a huge circular tray. The rice was garnished with dates, boiled eggs, and ghee, an oily substance made from butter. It tasted great! We sat cross-legged around the circular tray and ate with our right hand. Until I got the hang of it, I would sit on my left hand so that I would not inadvertently use it.

An important discovery. Gene investigates algae-encrusted ridges of submarine-cemented rock near Bahrain in 1966. The ridges explain the origin of enigmatic geological features, called tepees because of their resemblance to Native American dwellings.

You do not reach into the food with your left hand. The left hand is for sanitary purposes. You cleansed yourself using water and your left hand when squatting in the "thunder box." All dhows have a thunder box that hangs off the stern. There is no toilet paper. Muslims are clean, and they do not eat with their left hand! Hands and feet are also washed about five times a day before prayer.

When shark was not on the menu, it was hamour. I speared them as often as I could. Providing fish made me very popular with the crew. Hamour tastes better than shark!

Pearl Diving

During the thirty-day dhow expedition, Gerry and I collected hundreds of sediment and rock samples. It was an excellent experience, and it put me in contact with a pearl diver who would later invite me on pearl-diving trips. Here was the opportunity of a lifetime to document their diving with underwater photographs. I would write a story for *National Geographic*. I had already been writing a story about the catching of falcons. Not hunting with falcons, like you see and read about so often, but the catching of the falcons that would then be trained for hunting. (My story was never published, and to this day I have never seen an article describing how falcons are caught.)

National Geographic editors were amenable to the idea of a pearl-diving article, and I began sending them slides—about one hundred at a time. After each batch, the editors would make suggestions, such as, "Show us life in a pearling village." The next time it was, "Show us the children of the fishermen," and next it was, "Show the daily life on an Arab dhow."

To take all these pictures, I spent a lot of time in the small fishing village of Khor, north of Doha. Khor was where Rashid lived, and the people and their children already knew and trusted me. I always felt at home there. I took many excellent Kodachrome slides of children at Khor that—as the reader will see—would become important many years later. Unfortunately, I had to leave Qatar before finishing the story, and *National Geographic* returned all of my Kodachrome slides. Those color photos would gain importance years later as the children in the photographs became the leading citizens of Qatar. They would become the movers and shakers, and oil money would make them rich. Eventually, Pat and I would be brought back to celebrate our fiftieth wedding anniversary in Doha—all because of those Kodachrome slides. Who would have guessed!

When our tour of duty in Qatar ended, we returned to the Rijswijk laboratory for a six-month period of intensive report writing. Leaving Doha wasn't easy. Our servant Martine, a most pleasant fellow from Goa, had become part of the family. We hated to say good-bye to him and all the other friends and benefits that would be left behind.

Shinn family poses with house servant Martine (*left*) and Rashid (*right*) on their last day in Doha, 1967.

The work and report writing back in Holland were intensive and stressful, so much so that I developed a stomach ulcer. The Dutch doctor put me on a diet that also meant no alcohol. That was really tough.

Thanks to the design of Dutch toilets—you see the result of your labors—I noticed some unwanted intestinal guests. I informed the doctor, and he gave me two large worm pills. The first one flushed out about fifty roundworms and a very long tapeworm! I knew something about those worms. When I was a lab assistant teaching invertebrate zoology back at the University of Miami, I had dissected and memorized all their various parts. My ulcer cured quickly when the worms were gone, and I could once again relieve stress and enjoy fine wine.

It would be six months before we could return to the United States. During this period, our discoveries had begun to take on new meaning. The work on submarine cementation was important. Using the criteria developed by examining the large samples collected around Qatar, I was able to determine that the many small samples caught in the grabs and cores collected during the large Gulf-wide surveys were of the same origin. It was a lot like examining cuttings out in the oil patch. By mapping their distribution, we concluded that marine cementation had made a sheet of rock that covered about seventy thousand square kilometers of the Persian Gulf bottom. I would be carrying this message back to Shell University in Houston. It would not be a popular story back home!

Before leaving Holland, we had purchased a red 1964 Porsche convertible with black leather seats. It was a Super 90 and would zoom up to sixty miles per hour in second gear! With it we roamed the mountainous twisting roads of Belgium, made trips to Paris and Switzerland, and traveled the many back roads of France where there were no speed limits. On one trip we traveled with a French fellow named Jacques Gaouditz, who was married to a charming Swedish woman. We had been good friends in Doha where he worked for Schlumberger, the well-known international oil-field testing and well-logging company. Jacques had just purchased a brand-new Triumph sports car. With Jacques in the lead and translating, we threaded our way to a farmhouse retreat owned by his father. It was in the village of Boise Mangus, located near Normandy and about ninety miles west of Paris. World War II had bypassed this village, even though it was close to Normandy, and many had never seen an American. People in the village thought I was French! We learned during the trip that Jacques's father manufactured safety helmets for bikers and pilots. He was supplying aircraft helmets to the Israeli military during the "Seven Day" war with Egypt. Laws prevented shipments of war materials from France, so he transported them to Holland, from where they were shipped by air to Israel.

Another one of our many unforgettable adventures was in Italy. Mike Lloyd was attending a stable isotope meeting in Spoleto and was bringing his wife, Virginia. It was an opportunity we couldn't pass up, so Pat and I arranged to meet them there. First we flew to Rome and met up with Ed Purdy and his girlfriend. I had known Ed when he was a professor at Rice University sometime before he moved to visit Italy and study dolomite in the Alps. That night we shut down a basement nightclub dancing and playing conga drums with the small band. Just before sunrise, we sobered up with espresso at a sidewalk café.

Next day we rented a car for the drive to Spoleto. I had been watching the traffic in Rome. It looked and sounded like sheer pandemonium, and there were no stoplights! Could I survive it? Everyone behind the wheel drove like race drivers, but once in the traffic flow I found it quite easy. As wild and fast as they were, they were very good drivers. Italian gestures to other drivers were interesting—they included pulling on ear lobes or rubbing one's nose to question the other driver's heritage—and of course the one-finger salute. I played it safe and used none.

After circumnavigating the Coliseum, we were off through the hills to Spoleto. Mike and Virginia were waiting. I was not, and I am still not, an isotope chemist, but in an ancient ornate high-ceilinged room I listened patiently to presentations I did not fully understand. There was also a well-known American scientist there named Weber who was working on corals. What a contrast it was, listening to cutting-edge high-tech research in a baroque room hundreds of years old. Spoleto had once served as the French capital of Italy after it was conquered by Napoleon. By day, Pat and Virginia went sightseeing, and in the afternoons we watched elders play boccie ball in the park or roamed art galleries. We still have our inexpensive painting by a painter who is now quite famous. But, alas, it was about time to move on to another phase of my career.

4

Return to Stateside Shell

Battle with a Giant

The discovery of marine cementation put me at odds with the prevailing company view and a company giant. Our geologists were convinced that limestone formed only in the presence of freshwater! I would have to prove that this process was not peculiar to the Persian Gulf but had happened throughout Earth's history. Finding ancient examples didn't take long, but it remained controversial and soon we had two different camps contesting each other. The geologists at Royal Dutch Shell were quick to accept this new information. In fact, I'm not sure they had ever really converted to the freshwater idea that pervaded "Shell University" in the States. Who ever said scientists are gentle people?

Fortunately, we did eventually get over this bump, and various ways to distinguish the two different forms of limestone were established. Learning about the process and how to recognize its results led to many discoveries elsewhere. Major discoveries were made in coral reef settings around the world. Many were also made in deep cold-water settings.

The process is no longer controversial and is described in most modern textbooks on limestone origin. I suppose this process of rock formation would have eventually been discovered, but in this case it began ironically with the spearing of a fish for dinner. The seismic boosters also helped. So, did it change our thinking? Consider this.

The controversy over marine versus freshwater cementation within Shell became so intense that instead of sending me to the Shell Development

Research Laboratory in Houston, where I could begin searching for proof of marine cementation in the geologic record, management requested that I return to the Coral Gables Lab. They wanted me to work with Bob Dunham. By then, Bob Ginsburg had moved to Hopkins University and the lab had been put under Dunham's direction. Bob didn't believe in submarine cementation of carbonates and was not interested in looking at new evidence. Management hoped we could work together and iron out discrepancies. That was not to be!

I was still considered a bootstrap geologist. So, I returned to the Coral Gables Lab. Unfortunately, working closely with Bob Dunham would not happen. I would have to wait a little longer to search for evidence in the geologic record.

There I was, back at the Coral Gables Lab, which had changed somewhat. The move was very good in some ways. I will never forget the first day back at the lab. The first thing paleontologist George Herman did to welcome me was to get down on his knees and do a mock prayer. Next we all went across the street to the small restaurant where I had eaten for many years. The waitress looked at me and said in total deadpan, "You want your usual?" I had been gone three years!

Perhaps best of all, I was reunited with Ron Perkins and Paul Enos. Dunham had managed to obtain major funding to begin a regional coring project. They were drilling and coring the Pleistocene limestone under South Florida. This was Perkins's research project. He had developed ways to separate Pleistocene glacial and interglacial units based on exposure surfaces, basically brown caliche crusts—in essence, fossilized ancient soil horizons. Another significance of those caliche crusts would become apparent later.

Ron had also created a new stratigraphic nomenclature. He was not using conventional stratigraphic names that change from place to place. And he did not use the isotopic nomenclature that had been established for deepwater sediments. Instead he developed a simple "Q" system. Q stood for Quaternary, the most recent 1.8 million years of geologic time. He coded the units, from oldest to youngest, Q1 through Q5, with the youngest being Q5. These units could be traced throughout southeast Florida.

Meanwhile, Paul Enos began an extensive study of the Florida reef tract. He was using a high-resolution seismic device called a "mini sparker."[1]

Sparker-generated sound pulses penetrate sediment and reflect off the underlying Pleistocene limestone. The reflected sound is electronically converted to traces on a moving paper chart. It is much like a modern fish finder but much more powerful. In some cases, the sound pulses even penetrate deeper and reveal the same Pleistocene units that Perkins was drilling into and mapping.

To merge their data required drilling offshore. In one instance, a drilling truck was driven onto a World War II–era landing craft—similar to the salvage barge I had worked on years earlier—and unloaded on Little Molasses Island in the Florida Keys. (During my youth, Little Molasses Island supported some vegetation, but that changed when the island was swept away by Hurricane Donna in 1960; it is now a partially submerged pile of coral rubble five miles off Key Largo.) Gray Multer and Ed Hoffmeister had accomplished a similar feat a few years earlier but on a different reef. Drilling had to be completed quickly before any high winds or tides arrived. With great difficulty, Perkins and Enos core-drilled down to the Q3 unit. To reach it they had to penetrate forty feet of loose coral rubble, the most difficult material to core.

Perkins and Enos had done a monumental job, and of course the information was a company secret. They prepared company reports with the aim of helping geologists out in the oil patch to apply some new principles and better understand the stratigraphy of older subsurface rocks being explored. Eventually, some ten years later, their reports were released and published in 1977 as GSA Memoir 147 (Geological Society of America). They made many discoveries, but the reports do not reveal the exciting adventures that accompanied the fieldwork. Few geologists have such experiences today.

What had Bob Dunham been doing? Bob had initiated an elaborate vadose diagenesis project.[2] He wanted to verify the processes that would explain what he had described in his famous study of ancient limestones, which had greatly influenced how geologists thought about limestone formation. For chemical help he had enlisted Don Runnels, a geochemist from Shell University. Later, Evan Street, another Shell chemist, would come to do the required chemistry.

The centerpiece was a bulldozed pit in the Pleistocene limestone in South Miami. The hole, dug down to the permanent water table about eight feet below the surface, was located on the old World War II Richmond Airfield blimp base. Three vertical walls of the pit were instrumented with a variety of sensors. Chemical data derived from the sensors and various water analyses were to be compared with processes seen in the limestone using a microscope. Dunham was enamored with studying processes. "Process studies" were and remain popular in many fields of science. Unfortunately, geological processes are very slow; in fact, many are much too slow to occur during a human lifetime.

There was another problem. The Pleistocene limestone in South Florida almost completely lacks the kind of cement—fringes of needle-like crystals—that Dunham had so eloquently described from ancient limestone. The fringes of needle-like crystals are just one of the many glues that bind sedimentary particles together to form rock! Of course, it's not really a glue. It's just that the little needles interlock to form the bonding agent. Geologists call this process "lithification." This kind of bonding agent resembled what I had found growing in the rapidly forming rocks in the Persian Gulf. Dunham thought strongly that such cementing agents could only form in freshwater!

There were other problems. Basically, the Pleistocene rocks in South Florida are very different from most limestone in the geologic record. In short, the young Persian Gulf rocks with their submarine cements were much more like those encountered in ancient limestone than those in South Florida.

The reader can see my dilemma. Here I was, working for one of the most famous men in sedimentary geology, a man whose limestone classification scheme is universal. I was thoroughly intimidated and treaded very carefully. In spite of these problems, Bob Dunham and I got along very well, especially when we were discussing "fast cars, wine, women, and song." We just could not discuss the discovery I had made in the Persian Gulf, which seemed to undermine his most famous discoveries.

Bob was very intense. He had spent most of his professional career devoted to the interaction of freshwater and limestone. Along the way he had done some really unique experiments. One included hanging meat on a clothesline and examining it at regular intervals, similar to what forensic scientists do to learn how long a body has been dead. A more notable experiment was when he filled jars with those candy balls known as jawbreakers. He percolated various amounts of water through the jawbreakers to simulate the vadose environment. In the vadose zone, the water trick-ling downward to the water table tends to remain at the points of contact between grains. Chemical reactions take place at these points of contact. In the jawbreaker experiment, the water at the contact points caused the candies to merge into each other. Similar results can be seen in natural limestone where the grains interpenetrate each other. It was both an in-novative and instructive experiment. I could never find similar effects in submarine-cemented limestone. However, a few years later we would create those effects by experimentally compacting soft modern sediment.

Besides being intense about his science, Bob was also intense about other aspects of life. He had decided that some of mankind's problems are the result of being forced to abide by solar time. He was convinced that our mind and bodies were more tuned by the tides or lunar time. Female menstrual cycles, for example, are tuned to lunar cycles. Lunar time differs from solar time by about an hour each day and drifts out of sync with the solar day. We, like our clocks, are nevertheless locked into abiding by solar time, that is, night and day. Bob had correctly noted that our best mental acuity is during the few hours after awakening—don't we all feel like taking a nap in the afternoon?

For these reasons, Dunham had decided to work following lunar time. He would come to the office at 9 A.M., work for a few hours, go home for a nap, and return in the afternoon and work late into the night. The next morning he would start at 10 A.M. and repeat the process, with each day being about an hour later. Being a devoted scientist and passionate about this pursuit, he kept a running up-down graph of his sleep cycle on a roll of cash-register paper. We were all privy to what he was doing and knew the rules. We could cope with it, but it was a problem for management back in Houston. They never knew when to call and reach him in the office. This irregular schedule, lack of progress with the grand "process" experiment, and concerns about submarine cementation were probably the downfall of the Coral Gables office. Before the downfall, Dunham and I did collaborate on a field study to find submarine cementation in the Bahamas.

Three Stages of Discovery

In the early 1970s, Bill Taft at the University of South Florida had found rocky bottoms on the Bahama Bank that he claimed were examples of submarine lithification of modern sediment. Geologists at Shell—especially Dunham—had pooh-poohed his discovery. I remembered that episode quite well and went along with it until I made my discovery in the Persian Gulf. That discovery changed my view forever.

I also recalled that Ken Stockman and Bob Ginsburg told me about a place in the Bahamas where the bottom was rock hard and there was practically no sediment. The site, called Yellow Bank, happened to be where Taft had made his discovery. Yellow Bank is located south of New Providence Island. Stockman and Ginsburg had assumed the limestone there was Pleistocene, now submerged below water. It reminded me of our problem sampling on the Great Pearl Bank in the Persian Gulf. With that in mind, I convinced Dunham that we should go together and explore the area.

We went armed with an underwater vacuum device called an "airlift," picks, and hammers, and that old standby, dynamite! Digging in several places revealed that indeed there was rock at the surface but that a foot or more down the rock merged into loose carbonate sand. The old Pleistocene surface was a few feet farther down. What was interesting was that the large samples of rock that we collected looked almost exactly like the Pleistocene limestone on which the city of Miami is built. It was like the limestone in Dunham's instrumented pit, except that it contained the needle-like cement absent in his pit.

Bob Dunham was alarmingly faced with a rock that looked familiar but could not have formed under freshwater conditions, and he saw that the rock looked like the rock in his pit. The difference was that this rock was cemented with fibrous needle-like cement and also contained geopetal sediments.[3]

Before the Yellow Bank expedition was over, Dunham admitted that the rock did indeed look like the Miami Limestone but then asked, "But is it important?" "Is there much rock like this in the ancient geologic record?" Later I was to learn that this is the second stage in what is called "the three stages of discovery." Usually the three stages are (1) You are wrong and I can prove it! (2) You are right, but is it important? and finally, (3) You are right, but didn't we know this all the time? In an older version dating back to the time of the German explorer Alexander von Humboldt, stage three is, "Someone else gets the credit."

Management was becoming nervous about the Coral Gables office, and expenses were rising. By now Marlan Downey and Ray Thomasson were managing the headquarters laboratory in Houston—they decided it was time for a visit. When they arrived at the Coral Gables office, Dunham and I were brought into the same room with our bosses. Marlan and Ray began asking probing questions. It was the first time Bob and I had been forced to confront each other over the existence and implications of submarine cementation. I was uncomfortable, and Bob was even more uncomfortable but couldn't leave the room. In the past he would find a reason to leave the room, usually slamming the door as he left. In retrospect, this was the longest time during the past nine months that we had talked seriously about submarine cementation.

A few weeks after the visit, we were informed that the Coral Gables office was being shut down. We were all headed to Houston, and it became my job to work with the movers and pack up the entire office. All the equipment and the plastic-impregnated cores had to be moved. It was both painful and sad. Those cores were like a part of me. The only good news was that I would be going where I could examine ancient limestone to apply my findings, and I would be working with Mike Lloyd, who was doing isotope geochemistry. Mike had followed the debate and made sure that the samples I had sent from the Persian Gulf were quickly analyzed. He was into the issue and would now be my new mentor. I would also be working with Bob Walpole. Bob spoke fluent Spanish and knew his way around the significant outcrops in Mexico. He would help find examples of ancient marine cementation in the mountains of Mexico. And there was a new hire named Jerry Koch. The Shell University lab was a great place for me to do research and learn from the masters.

I would also be working with Paul Enos. Unfortunately for Shell, Ron Perkins took this move as an opportunity to acquire a teaching post at Duke University. Enos would wait a few years before making his break to teach at Binghamton, New York. Some years later, Paul moved back to his alma mater, the University of Kansas.

Ancient Reefs and a Different View

Earlier during our Florida Keys research, coral reefs had been avoided. We focused mainly on mud banks and tidal flats. They could be cored by pushing in a length of thin-walled tubing. It wasn't just because our coring method limited us to soft sediment. Geologists thought that most of the world's limestone and ancient reefs began their journey through time as piles of fine-grained lime mud. Ancient reefs do contain abundant lime mud. Ancient mollusks called rudists built enormous reefs during the Cretaceous—the time of the dinosaurs—that were thousands of feet thick. Fine-grained mud filled the spaces between the rudists and seemed to be analogous to the mud between the living finger corals that thrive on modern mud banks like Rodriguez Key bank. The ancient rudists and modern coral sticks seem to float in a matrix of lime mud when viewed in our plastic-impregnated cores of these banks. They did indeed resemble what geologists and paleontologists were describing from the ancient rocks. However, as serendipity would have it, reefs bashed open by ship groundings—or dynamite—showed there was also abundant mud within the coral reef framework.

Walter Adey and Ian MacIntyre at the Smithsonian Institution would later develop a diver-operated rock-coring device that clearly revealed the abundance of mud within coral reefs. With this new technology, many new observations were possible. Harold Hudson and I would eventually construct a similar coring device after I left Shell to join the U.S. Geological Survey. But there were other opportunities to observe the inside of reefs and see how much mud was hidden within.

It was Royal Dutch Shell policy to send people home for two months at company expense after two years of service in a foreign outpost. My home leave came while we were still living in Doha. Because my home was in Miami on the other side of the globe, this leave turned into an around-the-world trip of a lifetime. During our return to Miami, I arranged a side trip to visit my mentor Bob Ginsburg who was teaching a geology class at the Bermuda Biological Station. When I arrived I learned that Bob had obtained some special equipment that would allow me to play my "master blaster" role once again. Together, we broke open a so-called boiler reef using explosives. This time the explosive was TNT provided by the U.S. Navy base on Bermuda. How Bob talked them out of it I will never know! No one else I know could have done that. I had seen him in operation on the spur of the moment before. Bob could talk hungry dogs off a meat truck if he tried!

Boiler reefs occur in the most wave-affected parts of the Bermuda Platform—not a place where any fine-grained sediment like mud would be expected. This was a rough, wave-beaten environment exposed to the open Atlantic. Surprisingly, even here we found abundant internal mud, but this time much of that mud had already been cemented to form hard rock. The hardened mud formed geopetal surfaces in all the voids, which came in many sizes from millimeters to meters. This was where Bob invented the word "explosure" to describe our man-made underwater outcrops. One outcome of this adventure was that our Shell secret about submarine marine cementation was now out in the open. Soon the entire geological community would learn about this newly discovered rock-making process.

Now, why did I think it important to tell the reader about this? My first project back in Houston after shutting down the Coral Gables office was to examine about four hundred feet of core from the Lower Cretaceous reef trend in South Texas. The core came from about ten thousand feet below the surface and consisted almost wholly of rudists, an ice-cream-cone-shaped mollusk that built reefs like corals do today. During Cretaceous time, these organisms had constructed reefs as much as a thousand feet thick. They formed a twelve-hundred-mile-long reef trend—about the length of the Great Barrier Reef in Australia. The Cretaceous reef wraps around the Gulf of Mexico and even extends beneath Florida. The same reef is exposed in the mountains of eastern Mexico, where it can be studied in great detail. Where it is buried beneath Texas and Louisiana, parts of this ancient reef contain oil and gas.

When I began slicing and polishing the Cretaceous reef cores, it became evident that they contained a lot of what had once been lime mud. But this time I was armed with new information—I knew something I didn't know before. My experience developing criteria for recognizing marine cementation in the Persian Gulf, and what we had discovered in the Bermuda Boiler reefs, had shown what to look for. Sure enough, close examination revealed a thin rind of fibrous calcite around the rudist skeletons. The muddy sediment rested on this thin rind. It was clear that the mud had entered the reef after the rudists had been cemented together!

As discussed earlier, this was the time that such observations would have led geologists to believe the reef had been cemented in freshwater, that is, the mud would have been interpreted as having infiltrated the rock after initial freshwater cementation. In fact, the opposite was true. The lime mud had entered the rock during its formation, and it had all happened below the sea surface. It resembled what we had seen in the Bermuda boiler reefs.

That same process was later found to occur in modern coral reefs around the world, thanks to the development of diver-operated underwater coring devices. These processes were proven repeatedly by core borings and the old standby—dynamite! Previously we had studied sediment banks such as at Rodriguez and Tavernier Keys while ignoring the offshore coral reefs for the wrong reasons. Science often advances in unforeseen steps.

For me this was another example of how scientists—myself included—can be led astray following the ruling paradigms. As I like to say, running with the herd is safe, but sometimes the herd may be headed for a cliff. This is all in the past, but one should wonder what new cliffs lay ahead. I still feel fortunate to have speared that fish off the Zekrit jetty. It changed everything for me. Who would have guessed that such a discovery would have important implications for finding oil? When sediment is only partly cemented, it becomes a rock with pore space that can hold oil or gas. Partial cementation prevents compaction after burial that might otherwise destroy any porosity. Early marine cementation can be a good thing if it isn't too much.

Mud, Mud, Glorious Mud

"Mud, Mud, Glorious Mud" was the theme of a song written and recorded by Jerry Lucia of Shell University back in the 1960s. The song describes a field trip to the mud banks of Florida Bay, led by Bob Ginsburg. Mud is glorious because geologists had long recognized that a large percentage, perhaps more than half, of the planet's nonreef limestone had originally been fine-grained lime mud, and mud, because of its organic content, is more likely to be a source of oil than is sandy sediment.

That there was so much mud in the geologic record required us to conduct studies to determine its origin in modern environments. Early work on the Bahama Bank by Preston Cloud in the 1950s indicated that lime mud was being precipitated from seawater to form muddy clouds on the Bahama Bank, where they are known as "whitings." Once precipitated, the mud then settled to the bottom. These cloudy whitings form isolated patches and are surrounded by gin-clear water. Visibility in the whitings is limited to just a few inches.

After Preston Cloud's work, an isotope geochemist named Heinz Lowenstam (Bob Ginsburg's professor at the University of Chicago) noted that the tiny needle-shaped crystals of aragonite that form whitings resemble the aragonite needles that precipitate within the body of a small marine plant called *Penicillus*. *Penicillus* is just a few inches tall. It is called *Penicillus* because it looks similar to the microscopic mold *Penicillium notatum*, the source of the wonder drug. *Penicillus* grows on the seafloor over most of the Bahama Bank, as well as in Florida Bay and on the Florida reef track. Was this plant really the origin of modern lime mud on the Bahama Bank? If so, does it explain the origin of all the muddy limestone that formed during the long period of Earth history?

Put *Penicillus* plants in strong bleach, and the residue is aragonite crystals. This observation became the basis of a study headed by Ken Stockman, one of the original members of the small Coral Gables staff back in the 1950s. The idea was to stake out areas on the bottom and, using a frame with a grid of wire, photograph the precisely identical site every month. In some parts of Florida Bay the water was so turbid we used a Braille system. We kept track of each new plant as it appeared and died, and we could determine its life span.

Plants would come and go, allowing us to estimate the standing crop. By digesting representative plants in bleach and weighing the residue, we could keep track of how much aragonite was being produced in each plot. The results, eventually published in the *Journal of Sedimentary Petrology*, indicated that about one-third of the mud in Florida Bay could have been formed by these little algae.

At the time, we maintained that plants made the mud and then schools of fish stirred it up to form the cloudy whitings. Preston Cloud and others maintained that whitings were not formed that way but were caused by precipitation directly from the seawater. To some degree this controversy endures, and for many years Conrad Neumann and Lynton Land kept the fish-mud theory alive. Neumann and Land performed a more elaborate study of *Penicillus* based on the same idea as ours, but this time it was in the Bahamas, and their paper won the Society of Economic Paleontologists and Mineralogists annual best paper award.

Meanwhile, highly regarded geochemists such as Wally Broecker supported the fish-mud hypothesis, maintaining it is chemically impossible for aragonite to precipitate instantaneously from seawater. He strongly favored the *Penicillus*/fish-mud origin of whitings. I certainly went along with all this—that is, until I speared that fish in the Persian Gulf. If rock could form from precipitation of aragonite in seawater, when the best minds said it couldn't, then why not whitings? Our geochemists at Shell also said, "Direct precipitation of calcium carbonate from seawater to form rock was impossible." The pottery and glass in the rock had proven otherwise. Another problem was that *Penicillus* and other potential mud-making plants are not present in the Persian Gulf, yet there are abundant whitings there that look exactly like those on the Bahama Bank!

These observations gnawed at me and made me suspicious of the rules surrounding carbonate chemistry. For many years, chemists had pronounced that dolomite formation at Earth surface pressure and temperature was impossible, yet we found it forming in the Bahamas and Persian Gulf. An unexpected observation multiplied my suspicions.

While flying at a low altitude over rippled sand bottom in the Persian Gulf off the Trucial Coast, I spotted and photographed a vivid whiting floating above a sandy bottom. I knew right away that the sand could not be stirred up to form a whiting. That observation stuck with me and would serve as guide to another controversial discovery many years later. In the meantime, I would have to put on a coat and tie and be respectable. Whitings would have to wait.

Searching for Oil

After the Houston "Shell University" experience, it was a good time to go into the exploration part of Shell. I would go back to the oil patch, but this time it would be in New Orleans. I wanted to learn more and see if our new information about Cretaceous reefs could help find oil. Funding and emphasis at Shell Development Company were shifting. As we would often say, "A cold wind was blowing." I had a choice between Denver and New Orleans. I wrangled it so they would send me to Shell's New Orleans Onshore Exploration office, where I could begin mapping the Lower Cretaceous reef trend at a place where it passes thousands of feet beneath Louisiana. It was a new learning experience that was fun—and so was New Orleans! Fishing and diving around offshore oil rigs was excellent as well.

A seasoned geologist named Fred Strickland took me under his wing and showed me the ropes. At that time, most of the company's resources were focused on the Offshore Division, where exciting major discoveries were being made. The company was venturing into deeper and deeper waters, and motivation to drill offshore is different than onshore. Prospects offshore had to be drilled, or give up the lease. There was a time limit. Said another way, "Use it or lose it!" Whereas activity offshore was fast and furious, onshore drilling decisions could drag out for years. There was a lot more oil to be found offshore, and the Gulf of Mexico offshore region remains the United States' greatest source of domestic oil and gas. On the other hand, to get a well drilled onshore, the potential amount of oil had to be really spectacular. Otherwise, management wasn't interested. I learned much and did develop an onshore prospect, but it was clear that the company wasn't going to pursue it. I wasn't prepared to wait!

Australia and the Great Barrier Reef

While I was working on my onshore project, a call came from Australia. Management had decided to send me to Australia to testify before the Great Barrier Reef Commission, which was holding hearings on the widely publicized invasion of a large starfish called the "crown of thorns." *Acanthaster*, as this starfish is called, literally sucked the flesh out of coral skeletons. Why me? It was all about the results of the little study of staghorn coral growth, the study I had done on weekends back in Florida. What an opportunity! The year was 1972, and the environmental movement was in full swing and growing.

Environmentalists, predominantly biologists, claimed that the plague was caused by human activity, some kind of pollution. Geologists thought it was a natural cycle. One geologist had found old layers of starfish skeletons buried in reef sand and concluded that to be proof positive that the plague was not something new or man-made.

Because of the starfish threat, scientists from around the world converged on Australia to save the Great Barrier Reef. The U.S. government poured millions into research and containment. It was a great opportunity for budding reef scientists to see the Great Barrier Reef, and they went in droves.

Attempts to control the raging infestation involved divers injecting formaldehyde into live starfish. The method worked, but there were not enough divers to kill them all. Then, while the killing and the scientific battles raged, the starfish began dying of their own accord. The biological boom-and-bust cycle had run its course. It was all over, or was it?

Just as the plague was subsiding, a Japanese drilling ship arrived to drill on an oil concession owned by Broken Hill Propriety (BHP), an Australian mining and oil company. They had a concession near the Great Barrier Reef, and a drill ship that size could not be hidden. Environmental groups, already activated by the starfish invasion, feared the worst. "Save the Reef" bumper stickers were still available. Emotions were running high, and amid all the hubris, the Australian government was petitioned to stop the drilling. Drilling was put on hold, and a Great Barrier Reef Commission was created to evaluate the prospect of drilling and determine if it would do harm to the reef. Would corals be killed? If harmed, would they recover? How fast do corals grow? There are many, many species of staghorn coral on the barrier reef, and no one seemed to know how fast they grew! If the branches were broken, would they recover? Pretty basic stuff! I had broken enough coral to know it would keep growing. That was the limited state of knowledge in 1972.

Thanks to that little study I had published in 1966, BHP and Shell brought me to Australia to testify before the commission. Shell supported the trip because of the perceived precedent that might be created if environmentalists won the battle.

To get ready for the trip, I did some advanced reading and learned that Great Barrier Reef corals are exposed to the air at low tide for over an hour each day, yet they survive. Next I read about an experiment where a researcher wearing a backpack sprayer filled with crude oil sprayed the same exposed patch of corals every day for several weeks. None of the corals died, and the fish swimming in the pooled water around the corals paid no attention!

It was time for our annual vacation back to the Florida Keys, so I managed to obtain a five-gallon jug of Louisiana crude oil. Pat and I loaded up our equipment and the container of oil in our twenty-foot boat and trailer and drove to the Keys. On the first day, we found a suitable place in fifteen feet of water off Tavernier Key and wired foot-long lengths of live staghorn coral to two metal rods driven into the bottom. I also had brought along two small hemispherical clear plastic domes, the kind sold as skylights.

When all was set in place, I placed a small head coral under each dome and placed large clear plastic bags over the staghorn corals. It was tricky, but I managed to inject crude oil under one of the domes and into one of the plastic bags. Being lighter than water, the crude floated to the upper surface of the dome and did the same in the plastic bag. The other dome and bag were left without oil to serve as a control. The upper part of the head coral protruded into the oil, and about five inches of the staghorn coral protruded into the oil. The effect was immediate! The corals retracted their polyps, but something unanticipated also happened. The oil would not stick to the coral. Its mucus repelled the oil. I left the bag of oil and the plastic dome with oil over the corals for one and a half hours. I was trying to simulate the time that corals on the barrier reef might be exposed directly to a spill. When I removed the bags and the domes, I was certain the corals were doomed! Surprisingly, when we returned the next day they were alive and appeared well. At the end of our vacation fourteen days later we returned. All the corals were still alive and appeared healthy. I was ready for Australia and felt better about what I was supposed to do.

Regardless of the purpose of the Australia trip, it was a chance-of-a-lifetime experience for the Shinns. I had never seen the Great Barrier Reef and was especially eager. This was like going to Mecca! To prepare for the hearing, Pat and I were treated to a grand tour that took us most of the length of the Great Barrier Reef from Heron Island off Townsville to Green Island off Cairns. What a glorious trip it was! All was organized in view of future interrogation during which I could say, "Yes, I had seen major parts of the barrier reef." And we really did see the best parts, especially at Heron Island.

I had recently seen the crown-of-thorns starfish resting on a coral head at Heron Island on a television special back in the States. When I rented a boat at Heron Island, I asked the dock master, "Where are the starfish?" "We don't have any," he said. "But I saw the TV program about them being here." "Oh, Mate," he said, "those were brought here from a reef thirty kilometers away for the TV program."

The hearings were a new experience. In other courthouse rooms the judges wore white wigs just like in the old movies. Ours was a hearing, not a trial, however, so no wigs were worn. The hearings had been in progress for close to a year when I arrived. Written evidence lay in piles in front of the judge's high desk. There was close to a ton of paper piled up there.

The judge was seventy-five years old, and the company barrister told me the hearings were to be dragged out as long as possible because it was the judge's last case. On top of that, the government was paying the salaries for the prosecution barristers. There was no reason to get it over with in a hurry.

I took my place in the witness box and was interrogated for two and a half days. The main line of questioning related to the results of my paper, "Coral Growth Rate: An Environmental Indicator." I quickly learned much about how the law works from experiencing the process. The industry had hired the best of defense barristers, and he prepared me well for what was coming. The basic key was to answer yes or no when it would suffice, but under no circumstance should I provide an opening by saying more than necessary.

Having read the testimony of previous witnesses, I saw some good examples of what he was teaching me. Those who were against oil drilling often became emotional and said much more than needed. The new information would then lead to a new line of questioning. For example, one well-known underwater photographer talked about an observation he made near an oil spill. "When was that?" he was asked. "It was three days after our wedding," he replied. "And when was that?" he was asked. He did not remember the date. "So you can't remember the date of your marriage but you remember all the details of this oil spill you say you observed?"

That was a revealing example of what can happen when you get emotional. Giving testimony was easy but tiring. I soon learned to see where a line of questioning was leading from almost a mile away. They were setting traps, but I didn't take the bait. I answered the questions based on facts and with yes or no as much as possible. When it was all over I felt very pleased with my performance. With the barristers filling their wallets, the hearing went on after I left and lasted a total of two and a half years!

So why was I selected to do this in the first place? What caused it all was the support of a colleague, an Australian geologist named Bob Foster, who worked for the Shell Exploration office when we lived in Doha, Qatar. We had first met earlier at the Coral Gables office, and I had taken him diving when I was doing the coral-growth experiments. Later we were reunited when we both lived and worked in Doha. Bob was in the exploration part of Royal Dutch Shell while I was maintaining my small laboratory for the research arm of Royal Dutch. Bob and his wife, Jan, had been great companions on many camping trips around the Qatar Peninsula.

Foster eventually left Shell and returned to Australia, where he was hired by BHP. That convoluted series of international circumstances was how Bob knew about my interests and the staghorn coral growth-rate study. He had faith in my knowledge of coral reefs and had tipped off the BHP lawyers that they should bring me to Australia. What a strange turn of events it was! I didn't know it then, but the Australian adventure would lead to a major change in lifestyle and I was once again ready to take another fork in the road.

Learning the Business Side

One of my bosses and a great social friend and hunter/fisherman was a tall geologist/manager named Leighton Steward. As a student, Leighton played tight end on the football team at Southern Methodist in Dallas. He earned a master's in geology and after joining Shell was put in charge of Shell's deep-sea drilling project called "Eureka." *Eureka* was a drill ship outfitted to do test drilling in the deepest waters of the Gulf of Mexico. They were testing for oil related to salt domes even before technology was developed to produce at those depths. This was about the same time that the famous government-funded deep-sea drilling projects were gearing up. After the *Eureka* project, Leighton would find himself firmly in the middle of offshore oil exploration. Shell had been leading—and would continue to lead—the industry into new and deeper waters.

One day after I returned from Australia, Leighton came to my work area in New Orleans and said in his Texas accent, "With your biological background, you might have more running room at the head office. They are establishing an environmental affairs department there and could use your biological knowledge." I think he knew the company wasn't going to drill my onshore prospect. He also knew I was getting restless. I was ready for a new experience and jumped at his football analogy, "running room." By then, the company had elevated me to the title of senior geologist, but it was my biological knowledge they were after. The Australian experience had sealed my fate. Leighton thought my varied bootstrap background might be just what the company needed to solve environmental problems and deal with excitable, sometimes rabid, environmentalists. To sweeten the prospect, I knew by then that every transfer was accompanied by a salary increase. Such a move would allow us to upgrade to a better home, and besides, it was another opportunity to clean out the middle drawer of my desk. By then I had noted, like many others, that everywhere I went, the middle drawer quickly became overstuffed. Some things just crave a middle drawer! Moves usually solved that problem. So it was off again. We were headed back to Houston.

During the year in New Orleans I had purchased a speedy twenty-foot boat on a trailer. On weekends we fished and dived around the offshore oil platforms within reach of the Mississippi Delta, and Leighton along with several other geologists had become one of my fishing partners. That diving and fishing experience would pay off later because I would soon be diving around offshore platforms for more than just fun.

Another home to sell, and we were off again in search of a new one. Pat had earned a real estate license, so she handled the sale of our home, and by then we were old hands at moving. We found a two-story house in the better part of town off Memorial Drive that some jokingly called the "oil ghetto." It was a very upscale area. With the moving van leading the way, we drove directly to Houston and our new home. As soon as everything was unloaded, we were off to a party—all in the same day! That was a benefit of working for a good company. When they move people around, they pay moving and brokerage fees. Another benefit? There were always friends— fellow veteran geological gypsies wherever we moved.

Another moving story is worth the telling. Early on in my salvage-diving career, around 1958, I raised a five-hundred-pound cannon from the re- mains of a British ship on a reef in an area off the Florida Keys that would become Biscayne National Park. I had placed the cannon on a cement pad in front of our first home in Hialeah. It was aimed at, and intimidated, the neighbor across the street. That cannon would be moved many more times and always intimidated the neighbors. During our stay in the Netherlands and the Persian Gulf, Paul Enos and his wife, Carol, became our cannon keepers—it sat in their front yard. When we returned to the States, the cannon was recovered and moved several more times. The Queen, as we called the company, paid for each move. With each move it became a ritual to watch professional movers struggle with the quarter-ton monster. That cannon remains in the family and is presently intimidating our neighbors across the street!

Shell's head office was a different experience. I would be wearing a suit and tie to work and rode a bus to downtown Houston. I carried an umbrella and a briefcase and took the elevator with similar clones to reach the fifteenth floor of the fifty-five-floor One Shell Plaza building. It reminded me of the IBM stereotypes from the movies. It was definitely different from roaming the deserts and depths of the Persian Gulf. Our three boys attended the local high school, and Pat started something new—she took a two-year college course in interior design. After that she began decorating dentists' offices—of course, she was already an expert at decorating our various new homes.

We still had the boat, and there were abundant offshore rigs off Galveston for fishing and scuba diving. Although office life in a skyscraper was greatly different from lab work or the Florida Keys, I certainly learned much about what makes the world go round, so to speak. As a researcher, one was shielded from the business side of the business. One aspect of the work was very different. I could now talk industry problems—mainly environmental issues—with counterparts from other companies. At the same time, I was no longer privy to any of Shell's exploration and drilling secrets. Before this move, everything had been proprietary.

I was soon put on various American Petroleum Institute (API) committees. Through API membership, the industry began funding pollution-related research projects. I would find myself spending many days traveling to universities, monitoring the results of research funded by API. I spent much of that time driving up to Texas A&M to monitor progress on some API-funded research on oil toxicity. There were even trips to Hollywood to edit environmental movie scripts and commercial made-for-TV films. One day I flew to Los Angeles in the morning, viewed movie footage and script for accuracy, and flew back to Houston in time for dinner!

On another occasion I had a unique opportunity to spend ten days on Jacques Cousteau's boat *Calypso* serving as consultant and guide to some of my favorite areas in the Bahamas. Cousteau was making a TV special on blue holes. "Blue hole" is the local name for deep sinkholes now beneath the sea, and there are hundreds in the Bahamas. I would be experiencing this adventure with Bob Dill, one of the world's scuba-diving originals.

Aboard Captain Jacques Cousteau's vessel *Calypso* in 1973, Gene inspects the plastic resin cast of snapping-shrimp burrow made during an expedition to the Great Bahama Bank. This was the deepest crustacean burrow he had documented using his underwater resin-casting method.

The experience on the *Calypso* led to many speaking engagements at underwater film festivals, and because coral growth was a popular subject, I made regular presentations to diving audiences in places like Chicago and Boston.

Based on my diving experiences, I was elected president of the Houston Underwater Club, an exclusive jet-setting dive club composed of lawyers, doctors, and engineers. The club also hosted annual underwater film festivals and made frequent dive trips to the Caribbean. At one time there was even a television debate of sorts with Cousteau and his son Philippe, who was decidedly anti-oil and anti-American. To him, the oil platforms that hosted so much marine life only made it easier for people to catch the fish. At that time there were people and groups clamoring for old offshore oil platforms to be used as hotels, dive centers, you name it. These were all great ideas, but various legalities prevented the most innovative ideas from being used. The hitch was that if you owned the platform, one day in the future you were obliged to pay for its removal—and that would cost the owner millions of dollars!

There were many trips to offshore platforms where I would serve as guide for visiting politicians and environmental groups. The petroleum industry was being attacked from all quarters. It was also a time when Senator "Scoop" Jackson was calling for the creation of a federal oil company. Fishermen knew rigs were good places to fish, so I wrote articles about the marine life and regularly gave presentations to civic groups and dive clubs.

For various reasons, a popular Texas senator named "Babe" Swartz became interested in artificial reefs. He had access to surplus World War II liberty ships that could be sunk offshore to serve as artificial reefs for both diving and line fishing. Being president of the Houston Underwater Club made all these activities dovetail together. What I knew about coral reefs, artificial reefs, and fishing made good public relations for the company. One thing we could not do was coax government agencies to carry out research on fish and marine life around the rigs. The Man Under the Sea Technology (MUST) program and forerunner to what would become the National Undersea Research Program (NURP) in the National Oceanic and Atmospheric Administration (NOAA) was doing artificial-reef research. They were experimenting with reefs made of automobile tires and other materials, but they wouldn't go near a platform. I knew the inside story from Bob Dill, my diving buddy who was a key member of MUST. Because Senator Jackson was railing against the industry and calling for creation of a federal oil company, government agencies were savvy enough to know that the results of any studies they did might favor the industry position. The political climate was far from simple.

Together with Dana Larson, my counterpart at Humble Oil Company (they were in the process of changing their name to Exxon), we organized the first International Artificial Reef Conference. Of course we had our motives, but it didn't come as an assignment from higher up. We dreamed it up together and then convinced our management. We saw it as a way to show that offshore rigs were superior to sunken liberty ships, or to any specially built structures. As divers we already knew rigs made the best reefs, mainly because they provide habitat for a complete range of different fish species from the surface to the bottom. My position at Shell was full of diversity, but monitoring the work of other researchers and not being able to do it myself was bothering me. Dressing up, carrying an umbrella, and riding an elevator up fifteen floors every day was also unpleasant to me, and in the office I had to face angry phone calls from people asking questions like, "What are you going to do about the red wolf?" I didn't know a red wolf from a gray wolf. And there were encounters with what some called "little old ladies in tennis shoes" who always wanted to do good deeds and didn't like us. We were on the front lines and took the brunt of it all. It was distressing to learn that what most of the angry people really wanted was money for one project or another. In the end, it was all about money! I learned something else that I wasn't prepared to accept. I came from a culture where science was about truth. Your reputation was based on accurate description of nature. What I was to learn in that job was that there were PhD experts ready to give "expert" testimony on any side of a contentious issue. It was true in both government and industry. I found that difficult to accept, but then money speaks and will buy most anything but health. I became increasingly cynical of people's motives.

It was also frustrating that in a head-office staff position one has little authority to fund projects. You have a great expense account and jet all over the planet, but you can't promise anything to anyone or any well-meaning organization no matter how good the cause. That policy apparently hasn't changed. However, large companies have many doorways. I would tell people that if you are turned away at our door, then go to another. Eventually you might find support for your project in another part of the company. A little tip—if you are seeking funds from an oil company to support a project, don't go to the public-relations department if it is located in the company's head office. Approach another part of the company! Make friends!

I found it interesting that during the environmental movement, sightings of flying saucers seemed to plummet. "Maybe we should send up a satellite that would go beep-beep. Possibly those well-meaning folks would go back to star gazing and leave us alone." It was a joke of course—or was it?

Changing Times

A nagging concern was that I was fully aware that U.S. oil production had peaked around 1970. Twenty years earlier, Shell's own geologist, King

Hubbert, had predicted we would run out around 1970—and we did. I could only wonder, "How long would this job last in the face of declining oil production?" Would there be a need for this honest blue-eyed fellow in the three-piece suit to put on a good face for the company? Actually, I loved the company, but my experience in the head office had changed my outlook on the world. The only fault I saw was that I thought they were putting too many aging engineers in environmental positions. Those engineers couldn't relate to angry young activists. In fact many of these engineers projected the image of "big oil" and so-called fat cats, the very image that youthful activists disliked the most. Remember "Don't trust anyone over thirty"?

One rewarding experience while at the head office was visiting schools to give presentations. I especially remember making a presentation at a careers day event. The audience consisted of about seven hundred young people about to graduate from high school. What were they going to do? They were very worried about the environment and their future. Remembering my experience at the Great Barrier Reef hearings, I advised them to get two degrees: first, a degree in biology so they would have a science background, and then a degree in law. I said this because without this combination, they could not be effective in a courtroom. I had seen how biologists with a cause lacked facts and did not understand the rules of law. They performed badly at the Great Barrier Reef hearings, where they were made to look simply like irrational Marxists. Whether it was me or, more likely, it was plainly obvious to everyone, I am amazed at how many students have done exactly that today.

During this stage in life I was being constantly beat upon by angry citizens. That alone had a lot to do with deciding it was time to move on. I am very thin-skinned. Also, I had just turned forty. All of my friends said, "If you are going to make a life change, that's the age to do it," and don't forget, "Life begins at forty!" But what would I do? And where would I do it? Miami and the Florida Keys sure looked good from a skyscraper in downtown Houston!

Pat and I even considered buying a waterside motel on Key Largo. We had stayed there and met the owners, who wanted only $125,000 for the place. It was called the Florida Bay Motel; it came with a great little marina on the bay side of Key Largo, and I dreamed of setting it up as a base camp for visiting marine researchers and college field trips. I didn't have that kind of money, and my parents were wise enough to stay away from it. They had managed a motel once and wanted no part of it, especially my mother! Every time I drive by that motel on Key Largo, I am reminded that today that piece of property could not be purchased for less than five million dollars. What if? It was another fork in the road. Again, I took what turned out to be the right one.

The OPEC Oil Embargo and Disturbing Effects

The Arab oil embargoes came along, and they had a huge effect on my future. Because of my diving and fishing, one of my self-promoted projects was to have the company produce a fishing map for offshore Louisiana and Texas. I knew it was a good idea, because some enterprising fishermen had already obtained offshore lease maps and made blueprint copies that they sold in bait-and-tackle shops. Although really unprofessional and cheap looking, those maps did pinpoint oil-rig locations and provided the company names and lease block numbers. The little signs posted high up on the rigs were like a home address. A fisherman could look at the sign on the rig, check the map, and know exactly where he was. I had used these maps many times.

My plan was to produce a professional-looking map that would include color drawings of about thirty-two common fish species. It would show the kinds you could catch while fishing or diving under a rig. The reverse side of the map would explain the engineering differences between the various kinds of rigs. It would be good PR, not just for Shell but also for the entire industry, and they would be available for purchase, just like road maps, only at Shell service stations.

We even paid for a famous fish painter to do the illustrations, and Shell did a full-page advertisement in *Time* magazine featuring my statements about the kinds of fish you could catch, but the oil embargo wiped out my great idea. There was no need to advertise because there was so little gasoline available to sell. The company could sell all the gasoline it could get. Brand loyalty was not an issue as far as the public was concerned. In addition, the company lawyers pointed out that because other companies' rigs would be on the map, it would be like Shell advertising to have a picnic at an Exxon station.

There was yet another project zapped by the embargo. Twenty-five miles off Galveston Island sits Shell's Buccaneer gas field. It had become one of my favorite diving and fishing destinations. Gas at Buccaneer was dwindling, and they could hardly fulfill their contract with industrial users at nearby Freeport, Texas. It was about this time that I learned about how fish meal is made. Though fish meal has many uses, it provides the best way to stimulate chicken growth. It is made primarily from menhaden, which are caught in nets off Louisiana specifically for the fish-meal industry. Fish-meal-processing plants are smelly places—you don't want to live downwind from one. Here was my idea.

When shrimp boats catch shrimp, they haul in about ten pounds of so-called bycatch for every pound of shrimp. Those fish, usually called trash fish, could be converted to fish meal and fish oil, but instead they were, and still are, thrown overboard. Shrimp boats cannot carry enough ice to preserve the bycatch and take it to shore. Why not put a fish-meal plant offshore near where the shrimp boats fish? They might economically bring the bycatch to the platform for sale and processing. Foul odors were no problem out there.

I checked with the production engineers, and they assured me there would be enough natural gas at Buccaneer to power a fish-meal-processing plant for the next hundred years! Great!

Management liked the idea, and we began negotiating with a fish-meal-processing company. The prospects looked good. I was elated! But then the embargo drove the price of natural gas upward, and the engineers did what they called a "work over." They redrilled and teased more gas from the existing wells, and today Buccaneer field is still producing gas for various Freeport industries. It was another frustration caused by the embargo.

Another project that seemed made to order was to install an underwater chamber on the leg of an oil rig. It would allow visiting politicians and the like to actually descend about thirty feet and observe all the fish. I got the idea from two places. First, during the coral reef hearings in Australia we visited a fixed underwater chamber with windows. We could look out through portholes at all the colorful fish and coral. I watched tourist reactions—they loved it. Second, while doing a ten-day trip on the *Calypso*, I would climb down into a circular chamber at the bow and watch the fish and dolphins as we cruised along. It was a very simple but effective device. With this in mind, I proposed the idea to management—not at the head office but to those in charge of offshore drilling. I drew pictures and described how to include a hand-operated windshield-wiper-like device to keep the portholes from being fouled. Some months later, I heard Offshore Division had actually installed the device. I had to pull some strings to get offshore and inspect the chamber.

Instead of climbing rungs, the engineers had outdone themselves. They had installed a small electric elevator so three people at a time could descend. The problem was that it had been installed down current from the pipe that discharged a cloud of drilling mud. Mostly what you saw were clouds of mud. Not exactly what you want to show visiting environmentalists.

Then there was the problem of fouling on the viewing port. There was no device for cleaning the portholes. Through various sources, I learned that the plan was to have a professional diver clean the ports about every two weeks—at great expense, I might add. It irked me that my simple idea had turned into a monster. It wasn't long before they stopped cleaning the portholes. They fouled in about three days! I could have told them that would happen.

The underwater-chamber experiment led to yet another idea. In 1935 a man in the Florida Keys had installed a vertical chamber with portholes to view fish and coral at one of the best coral reefs in the Keys. It was called the Sea Aquarium, and the chamber was subsequently knocked over in twenty feet of water during the great Labor Day Hurricane of 1935. When I first began diving I had seen the chamber lying on the bottom. Later, a scrap-iron salvor whom I knew recovered it and sold it at the yard up the Miami River as scrap iron. With the old Sea Aquarium in mind, I proposed that Shell construct a similar but much larger and improved device. It could be installed on the protected side of that same reef where I did the growth-rate studies that led to my trip to Australia. I had an artist do a sketch based on a viewing chamber already operating in Japan.

Management, I'm sure, viewed this as another crackpot idea. Imagine what a public-relations coup this would have been for Shell! In those days, the Marine Sanctuary system was not yet created, so there would have been no legal reason to stop the project. Such a chamber would give nonswimmers a chance to view the best coral reef in the Florida Keys, and they wouldn't even have to get wet! Today, hundreds of snorkelers are transported in so-called cattle boats out to my favorite reef almost every weekend while nonswimmers have to settle for going out on glass-bottom boats. These negative experiences were starting to add up, and I was now really tired of wearing a suit and tie every day. I could sense a fork in the road ahead.

Shedding My Shell

I had heard through the grapevine about a new research organization in Florida called the Harbor Branch Foundation, whose facilities were located a few miles north of Fort Pierce. That might be a neat place to go, and besides, my parents had retired to nearby Port St. Lucie.

One of the great advantages of the head office job was that I had incredible access to information. For example, I knew that Harbor Branch was interested in coral reef research. I knew they had received federal funding for a huge baseline study of the Indian River Estuary. So-called baseline studies were becoming very popular. They were studies to map and describe everything in selected areas, so that if the environment were degraded by some activity, it would be easier to quantify the change. The data also provided information for Environmental Impact Statements (EISs) that were required for any project that might have an impact on the environment. Many organizations specialized in doing EISs mainly for government agencies.

What was unusual about the Harbor Branch baseline study was that, for unexplained reasons, it included a narrow strip extending twelve miles out into the Straits of Florida. That was well away from the Indian River Estuary. What was that about? With resources available, I was able to investigate further. As I suspected, there was something afoot, so I checked with many sources.

A state employee who had also once been a technician for Bob Ginsburg at the Shell Coral Gables Laboratory had told me the state had brought charges against Harbor Branch for infringement of the mangrove shoreline. That was interesting. Then I read Platts *Oilgram*, an industry-wide newsletter that crossed my desk every day. It said that Seward Johnson, the power behind Harbor Branch, was negotiating with a rich rancher a few miles inland of the laboratory. There were plans to build an oil refinery on his ranch. Crude oil could be transferred to the refinery via pipeline from tankers offshore using what was called a single-point mooring. So I assumed that the offshore strip to be studied was the likely site for the pipeline and single-point mooring system.

In addition to all this, my father's next-door neighbor in Port St. Lucie was president of a local bank. He did a little checking and confirmed that Harbor Branch was indeed solvent. There were no financial problems. Then, through a chance meeting, I met an older gentleman in Houston who called himself the family historian for the Johnson family. I indicated my desire to work at Harbor Branch, whereupon he took me aside to give advice. His principal advice was, "Insist on a contract and expect them to haggle with you down to the last penny. Those folks did not get rich giving their money away." That was good advice. I found out for myself how right he was.

What I didn't know was that the Woods Hole Oceanographic Institute and Biological Laboratory and the Smithsonian Institution were trying to get a foothold in the Harbor Branch organization. Some weeks later I met people who jokingly referred to the place as Harbor Hole. So, what was the Harbor Branch Foundation?

As near as I could learn, Seward Johnson, of Johnson and Johnson medical, had met Edward Link, inventor of the famous Link Trainer—used to train World War II pilots—on a trip in Jamaica. The story was that their boats were in the same marina. Ed Link was a well-known treasure hunter and had been salvaging the sunken city of Port Royal, Jamaica. Another story was that Seward was investigating native remedies, especially aphrodisiacs. They met, became friends, and came up with the idea of a foundation that would focus on development of a new kind of submarine and jointly look for new drugs in the sea. That alliance actually led to development of the famous Johnson *Sea-Link* submersible.

The Johnson *Sea-Link* features a unique acrylic bubble-shaped forward hull for the pilots. It had a separate aluminum hull in the rear that would allow divers to lock in or out while submerged. Divers could leave and return to the submarine while at depth. There was also the prospect of Johnson and Johnson researchers developing new drugs from the sea. I can only guess that a tax break would aid the effort, especially if it supported coral reef research and other scientific research that required the submarine. By 1974 the *Sea-Link* was in full operation. The proposed oil refinery never came into being, probably because of the oil embargoes.

Interview, Almost Fatal

Harbor Branch had experienced a significant turnover of research directors and was then looking for a new candidate. I also knew they had a policy of hiring only PhDs. I probably wouldn't fit the bill. Nevertheless, I wrote up a list of potential coral reef research projects and mailed them off. The letter led to an invitation. Come to Florida for an interview. That was ideal, because I could stay with my parents a few miles down the road.

The interview was interesting in a number of ways. I had the opportunity to meet Mary Rice, who was one of the original researchers. Her specialty was a truly academic study of a special family of worms. Others I met were investigating crabs and other organisms known to inhabit the Indian River Lagoon. They were heavily into the baseline-study concept that was so popular back then. As mentioned, these comprehensive studies were needed to prepare an EIS. All new major projects, including an offshore tanker terminal or any shoreline development, would certainly need one. The organization was thinking ahead, but the researchers were unaware. They were just doing interesting science. Interestingly, the government was funding the study.

At that time, most of the work at Harbor Branch was being conducted in a laboratory on a floating barge in the artificial canal that the Harbor Branch Foundation had excavated. There was also a two-story cement building. It housed the submarine assembly building that was on the other side of the blind-end canal.

As fate would have it, Ed Link and the submersible were away on the day I arrived, but I did have an interesting lunch with Seward Johnson and Adair Feldman. It was at the Hilltop Café on U.S. 1. Seward was a likable, unpretentious gentleman dressed in old clothes. He wore a floppy hat. Conversation was pleasant. He was a friend of another elderly gentleman in the area named Ralph Evenrude, the man behind Evenrude outboard motors. My father had met Evenrude, and the Evenrude test facility was a stone's throw from the Harbor Branch facility.

After lunch I met with Marilyn Link, Ed Link's sister. She appeared to be the business person in charge. She showed me some of their stock holdings. I knew nothing about stocks but acted interested. As predicted by the gentleman back in Houston, they had no interest in providing contracts for their employees. "We are all like family here," she explained. Marilyn was pleasant and later invited me to drive down to Key West for the weekend where I could go out to the boat and meet her brother. They were conducting submarine-diving operations. I had heard that scientists found him difficult and decided instead to visit with my parents. It was a good decision. I took the right fork this time. If I hadn't made that choice, the reader might not be reading this story!

Marilyn turned me over to chief scientist Dr. Adair Feldman, who was on loan from the Smithsonian Institution. He was a very pleasant individual. We had bonded when we lunched together. I made the mistake of asking some pointed questions about the proposed oil refinery, a potential oil pipeline, and also the state's lawsuit over the mangrove shoreline. He apparently knew nothing about these matters. We finished our conversation on a Friday afternoon, and I drove south to spend the weekend with my parents in Port St. Lucie.

While visiting my parents, I heard the news that the *Sea-Link* was entangled in a cable on a sunken Navy destroyer in over two hundred feet of water. It was near Key West. Ed Link's son and another fellow were in the aluminum lockout chamber. Apparently because of the cold water at that depth, the aluminum chamber became so cold that the chemicals used to scrub carbon dioxide from the breathing air became inefficient. They both died from carbon dioxide buildup and hypothermia.

Had they made an early decision to lock out, they might have made it to the surface. I say this because another good friend, Dr. Richard Slater, had made such a lockout assent from a disabled submersible in 280 feet of water off California. He lived even though he was unconscious when he reached the surface. He recovered and is alive and healthy today.

Those in the *Sea-Link* lockout chamber remained in the chamber too long, believing help was on the way. There was plenty of blame to spread around. The Navy would not allow others to make rescue attempts. A friend back in Houston who headed an oil-services diving company, Oceaneering International, had his divers on standby to fly over, but they weren't invited.

The sub was stuck at the depth this company normally operated. The Navy operated either at much greater or much shallower depths. They apparently were not equipped for this intermediate depth. It would have been duck soup for divers in the offshore oil fields.

Fortunately, the pilot and an ichthyologist named Bob Meek were in the more insulated acrylic forward-control bubble. They survived long enough for the sub to be recovered. An independent operator using a grappling hook literally jerked them loose. He had a video camera strapped to the grappling hook for guidance. Bob Meek told me years later that they too would have expired had the rescue come fifteen minutes later. Because the two in the rear chamber were really there to serve as ballast, I feared that had I accepted the invitation, I might have been dead weight.

The accident resulted in significant operational modifications. Before the accident, the automatic snap hook for sub recovery was on the submarine. It had snapped shut onto a cable attached to the destroyer. The system was soon changed, and the hook was placed instead on the recovery crane.

Years later I did a mission in the *Sea-Link*, and it was very safe—I was also in the forward bubble. Later I became friends with Bob Meek, the biologist who survived the ordeal in the forward bubble. He had formed a company in Santa Barbara called EcoMar and specialized in studies for petroleum companies in California. Meek also developed the commercial culture of edible marine mussels on offshore platforms off Santa Barbara. They were especially good because they did not contain grit or other contaminants often common to mussels grown near the polluted shore. He battled long and hard with authorities, who were against anything associated with offshore rigs—a not uncommon sentiment in California.

A few weeks after my visit to Harbor Branch, I received a letter from Marilyn Link. She said they felt it was better that I stay with the secure job I already had. I suspected I had asked too many questions—and besides, I had insisted on a contract.

In response to the rejection letter I suggested a fellow whom they might hire as a coral reef researcher. He was Dr. Philip Dustan, an innovative marine biologist in the process of graduating from Stony Brook University, which is part of the New York University system. We had met when I visited Stony Brook to give a talk. Phil was accepted and set up a field station for Harbor Branch on Key Largo at Pennekamp State Park. He had an enviable position for several years, and we would cross paths and collaborate many times in the future. Nevertheless, things for me would change for the better. A job was opening up that would lead to the most productive thirty-one years of my life. Life does begin at forty if you're on the right road!

5

The U.S. Geological Survey

All About Pete

Cars, all shapes and sizes, were backed up for blocks. Reaching the gas pumps could take hours. That's the way it was during the height of the 1970s oil embargoes. The American dream had become a nightmare, and Congress responded by literally throwing money at the U.S. Geological Survey (USGS). The nation needed a quick fix! Congress needed information, and the United States badly needed to expand domestic oil production. How bad was it? How much oil was left in the United States? We needed to know! The USGS had what could be called a sleepy Oil and Gas Branch based in Denver, Colorado, that had existed for years but had long ago fallen by the wayside. Apparently, they mainly shuffled a lot of paper and kept records.

Then along came one of my longtime friends from Shell. I had been his diving technician when he did his six-month assignment at the old Coral Gables Laboratory. We had been good friends before I went off to the Persian Gulf, and he had been hired away from Shell by the USGS. His mission? Go to Denver and revive the sleepy Oil and Gas Branch. His name was Peter R. Rose, a down-to-earth grassroots Texan.

When we first met in 1961, Pete had a master's degree from the University of Texas. Although his graduate work was in carbonate stratig-raphy[1]—he had studied under Bob Folk, a famous "guru" of carbonate sedimentology—Shell had put him to work in their Houston office as a micropaleontologist, or "bug-picker," to use the oil-patch vernacular. He worked with Tertiary rocks of the Gulf Coast, where he studied tiny fossil foraminifera, the "bugs." He came to Coral Gables to study the living ones in Florida Bay. Maybe seeing how the live ones live would help us better understand the dead ones that were so important, in turn, to understanding the geologic age of rocks in Texas. Shell was making a big play in the Lower Cretaceous Edwards Reef beneath the Gulf Coastal Plain, so perhaps the study would help them find more oil and gas.

Pete had taken a winding road full of forks to reach the Oil and Gas Branch. After six years with Shell, he went back to school and earned a PhD from the University of Texas. His dissertation study was on the regional stratigraphy of the Edwards Limestone. The subsurface part of his study drew on his Shell work when he was stationed in Corpus Christi, following his Coral Gables tour, and the surface-outcrop work came from the Edwards Plateau of west-central Texas, literally on his native soil where his family ranch still sits.

In the course of mapping, Pete had found some distinctive thin gray limestone beds. They had been bored on top by rock-boring clams. These strata were more than one hundred million years old but looked exactly like the marine beds forming in the Persian Gulf—the ones containing the glass bottles, nuts and bolts, pottery, and so forth. What was interesting about Pete's bored beds was that many of the borings—and their sediment fillings—were themselves penetrated by later borings. Some borings and their fillings had been bored many times. We had found the very same features in the modern Persian Gulf submarine beds!

Needless to say, we were pulled together once again. Conventional Shell dogma held that such beds could only become cemented during periods of exposure, when the sea left them high and dry. Following the Dunham and Company dogma discussed earlier, Pete's beds would have had to have been repeatedly exposed, then reimmersed in the sea and bored by the clams, in some cases many times. The sea would have been like a yo-yo! My discoveries in the Persian Gulf showed that such complications were not necessary. The layers could have formed in the same salty water the boring clams inhabited. Pete and I were fighting the same scientific battle.

Pete had rejoined Shell in 1969, after a year of teaching at SUNY Stony Brook. The Vietnam War was on. It was the height of student rebellions. "Don't trust anyone over thirty" was their slogan. Stony Brook experienced predawn drug raids and the faculty did not seem to care and the administration backed down under pressure. Pete, who came from an old Texas ranching family, had had enough of that liberal environment, so he went back to Shell. It was during the first oil embargo that the USGS came looking for him, and that was how he became chief of the Oil and Gas Branch, a move that shocked conventional USGS employees.

While at Shell, Pete had become involved in recruiting for the company. He had visited many universities and knew the locations of good students who would be job hunting. With his company-recruiting background and knowledge of where outstanding researchers could be found, Pete hit the ground running. He really would reinvigorate the sleepy Oil and Gas Branch.

I had heard about this and on a hunch made a phone call. "Pete, it sounds like you are reinventing Shell University within the USGS. Don't you need a field office in Coral Gables to study modern carbonates?" Pete laughed, and we had a pleasant conversation. Actually, it wasn't that pleasant. His fourteen-year-old daughter had suffered a terrible horseback accident. She would be permanently paralyzed from the waist down and was just beginning rehabilitation. I could tell that Pete was suffering.

The very next day Pete called back, "I talked to my colleagues, and they liked the idea. When do you want to start?" Two weeks later, I was in South Miami looking for a home! This was a really big fork in my life road, and I took the correct one. It was another example of unexpected good timing—and connecting with capable people who recognized latent potential.

Securing the necessary funding to get a new venture started is never easy, but serendipity came my way yet again. One of the main charges of the Oil and Gas Branch was resource assessment—how much oil and gas did the United States have left? That wasn't really my specialty, but fortunately there was enough funding for what the USGS called "topical studies," which were studies that might aid those doing the assessments. Resource geologists needed to know more about carbonate rocks, and I still had that list of ideas that I had put together for Harbor Branch.

Our first goal was to recruit a small group that would do research on carbonate sedimentation. Pete was pushing rapid publication of results, in what were called USGS Open-File Reports that are made available to the public within a few months of completion of the research. The old-line USGS folks had been used to leisurely publication of their work. Many disliked Pete's enthusiasm for Open-File Reports. When Pete and I had

worked for Shell, reports were secret, often taking years before being released for publication in scientific journals. Now we could publish at will.

Our goal was to get information out in the open literature so that small
companies and independents could use it quickly. Although the deeper,
expensive, offshore ventures would remain the domain of the majors, small
companies and independents were accounting for about half the oil then
being found in the United States. We were to help the little guys who didn't
have high-powered research departments. Of course, anything we did was
there for the big boys as well.

What was most rewarding and encouraging was Pete's charge, "I want
you to do what you think is important and publish in the open literature."
It was sweet music and, like the song, "Who could ask for anything more?"
We would establish a field station in Miami with headquarters in Denver. It
was reminiscent of my years in the Persian Gulf, when I was following my
research instincts and headquarters was far away back in Holland. Pete and
I agreed from the start that this project would last only five years. It lasted
fifteen!

There was one other factor to consider. There were people in Congress,
especially Senator Scoop Jackson, who were calling for a federal oil company. The oil industry was very uneasy. They imagined they might soon be
competing with a government-run oil company right here in the United
States—shades of the WPA and TVA from the Depression era! I didn't like
that idea, and neither did Pete. Some in the industry, however, thought we
might. They feared the USGS could be the tip of the iceberg, the germinat-
ing seed of "Federal Oil." Many remained skeptical—including Bob Nanz,
Shell's executive VP for exploration. It didn't help that Pete had recruited a
half-dozen seasoned geologists from Shell to staff the revived Oil and Gas
Branch.

Fisher Island Station and Building a Team

Finding office space was the first chore. I began negotiating with Harris
B. Stewart. "Stew," as he was called, headed up the NOAA Atlantic Meteorological and Oceanic Laboratory on Virginia Key. It was right across the
street from the University of Miami Marine Lab where I had first worked in
1958. I could look over and see the building under which I had crawled with

the union plumber years earlier. Stew provided a temporary office, but we felt like strangers. We were small fish in a big pond—and a different branch of government owned the pond.

One morning, Pete called and said, "There is a young PhD named Bob Halley who applied for a job. He does carbonates. Would he fit in down there in Miami?" You bet! I had met Bob about a year earlier, when he had been a postdoc with Paul Enos. I already knew Paul liked Bob and thought he was innovative, exactly what Pete and I wanted.

A few weeks later, Bob and his wife, Barbara, drove to Miami in a worn-out Ford Bronco. They had a small child with them and another on the way. Before Bob arrived, Pete and I had also recruited Harold Hudson, my lifelong buddy and fellow diver who was working as a biologist. He was next door at the National Marine Fisheries Laboratory (NMFS). He had always wanted to work on corals, but NMFS, which had no interest in coral or anything not edible, always told him, "You don't eat coral; you eat fish and shrimp." Harold was also the NMFS diving officer at their Virginia Key facility. He knew all the government regulations for diving—another plus, as our work would revolve mainly around field research that required diving.

Soon, Cesare Emiliani called from the University of Miami Marine Lab. His National Science Foundation (NSF) money was dwindling. He had this young assistant who was an expert with foraminifera. "She has a photographic memory for recognizing these little bugs," he said. I knew Pete had a fondness for bugs, which were the same critters that Pete Rose had studied during his six-month assignment at the old Shell Laboratory. Bugs were important to the oil business. However, I was also looking for a secretary.

Cesare said this lady needed a steady job. Her micropaleontologist husband, Lou Lidz, had recently succumbed to leukemia, their eight-year-old daughter had succumbed seven months later during exploratory surgery, and she was raising a six-year-old daughter. Her name was Barbara Lidz, and she arrived for an interview the next day with her portfolio of published research papers and published, hand-drawn illustrations of microfossils. "Yes, yes, very impressive," I said, "but can you type?" Yes she could! There was only one problem. She specialized in planktic foraminifera. Paleontologists in the oil industry studied benthic foraminifera. Benthic species live on the seafloor and are used to tell relative depths below sea level at time of deposition of ancient marine sediments. Planktic foraminifera live in the water column, sink to the seafloor upon death, and have distinct geologic-age ranges that allow geologists to determine the age of the rock being drilled in which they are found. Barbara agreed to make the switch and incidentally in her spare time she could take care of office business. Wow! Did she ever!

Barbara had been editor for the Miami Geological Society. In fact, she had held every elective office in the society, including president. She could edit, spell, type like a fiend, and, oh yes, she could draft, manage our budget, and do her own research. Computers were still way in the future, so we were using typewriters, zipatone letters, and chartpak and zipatone shading for illustrations in our reports and papers. We had only recently graduated from Leroy pen sets. They went out with the slide rule.

As a team, we could do everything including photography, diving, X-radiography, and outboard-motor repairs. We could saw rocks, impregnate sediment with plastic, make and analyze petrographic thin sections, and soon we would be drillers. We were more like a small company than part of a big federal program. Best of all, we were having fun! We took pride in what we could accomplish with so little. We also learned to work the system. Our working motto would soon become "'Tis easier to seek forgiveness than to seek permission." We made a sign that proclaimed we were the "Federal Skunk Works." Thirty-six years later, that sign adorns the wall over Barbara's present USGS office door.

We had the basic four-person team but needed a larger work space. Within the large new NOAA building we were lost, and we all shared one office! Soon we heard about a place over on the next island. It was called Fisher Island, named after Carl Fisher, the developer and builder of the Indianapolis Speedway. On the island was a government quarantine station that had been built in the late 1920s along the shore of Government Cut, the entry into the Port of Miami. The quarantine station was established back when commerce and people arrived by ship. Quarantine operations had long been shifted to the Miami Airport, so the empty station, situated on fourteen acres, had been given to the University of Miami on a thirty-year lease. Hmmm, I thought—I might be retiring about then!

Facilities consisted of four Spanish-style four-bedroom, two-bath homes and a larger administration building. There was also a boathouse, a machine shop, and a university caretaker who ran a small ferryboat across Government Cut to a parking lot at the south end of Miami Beach. He ran the boat every hour on the hour when needed. There was no bridge, and it remains that way today.

For the most part, the island was covered with Australian pines that harbored voracious mosquitoes. At times it seemed they could carry you from the boathouse to the office door! Back in the mosquito-infested woods was an old stone mausoleum. Over the entrance it said "FISHER." There were four empty crypts inside. Carl Fisher had not been interred there, although clearly, someone had been thinking ahead.

On the southeast corner of the island was an elaborate abandoned estate. It was a large, spooky place, complete with greenhouse, pool, seaplane hangar, and a seaplane ramp. It had been home to a reclusive member of the Rockefeller family. After Rockefeller it had been home to Gar Wood, who had made his fortune manufacturing hydraulic rams and pumps. He also made plywood boats for the U.S. Navy during World War II. He had even installed a single-lane bowling alley with automated pinsetter.

After Gar Wood, Bebe Rebozo acquired the estate. In fact, Rebozo owned most of the island property outside the quarantine station acreage. Belcher Oil owned the remainder. They had huge oil tanks for refueling ships and barges that supplied the electric power plant in south Biscayne Bay. The reader will probably know that Rebozo was a close friend to President Richard Nixon. Rumor had it that he and Nixon intended to build a bridge from the mainland to the island so they could develop the property. A bridge would have made the mosquito-infested island into very expensive real estate. Developed, the island would be worth many millions. Years later it actually was developed, and presently it supports an exclusive resort, but there still is no bridge. It retains its exclusivity by using a car ferry. The ultimate gated community!

Barbara Lidz had known Bebe Rebozo from when she lived on Key Biscayne. She banked at Rebozo's bank. In fact, most everyone on the island knew Bebe, who was a very likable person. Every Friday, on payday, he personally popped corn in the bank lobby and handed bags to patrons standing in line for a cashier. My father had also known Rebozo in the late 1940s when EK was the FAA branch chief at Miami Airport. Back then, Bebe ran a small tire-recapping business and sold tires to the FAA. Interestingly, I later learned that Nixon had championed the idea of tire recapping as part of the war effort. Before long, I would also know Bebe as well as his brother Mike, who was in charge of the property on Fisher Island.

Another geologist had beaten us to Fisher Island. He occupied the large administration building. It was none other than my former boss and mentor Bob Ginsburg—the man who had converted this diving drummer boy into a geologist. Bob had left Shell to teach at Johns Hopkins University when I went off to join Royal Dutch Shell in Holland. He had also taught summer courses at the Bermuda Biological Station. The Baltimore climate and stiff academic environment at Johns Hopkins had forced him and his wife, Helen, back to warm and sunny Miami. Bob had always said he wanted a one-elephant circus. He ran a small circus at Coral Gables when I went to work for him. He said there were too many elephants at Johns Hopkins and the circus was too big. Now he had his small circus, just like the one at the old Coral Gables office, but this one was surrounded by water! The question was, could the island support two elephants?

Cesare Emiliani, Bob's former classmate at the University of Chicago, had introduced him to the school's new prize—the empty quarantine station and its large administration building. A University of Miami microbiology project had operated there earlier and then ran out of money, leaving excellent laboratory space behind. We wanted it badly.

Ginsburg was now teaching graduate classes. He had also established and ran an Industrial Associates Program funded by the petroleum industry. At times his program was funded by as many as a dozen different oil companies! Like the old Coral Gables office, his program was a magnetic focus of intellectual activity. They came from around the world. They were the leaders in carbonate sedimentology.

The tricky problem for us was that the petroleum industry feared a federal oil company was brewing. Might the USGS become Federal Oil? Would we become competition? The movement in Congress to establish a national oil company was still viable, so some saw us as a potential enemy, especially if we published in the open literature. Bob's work was for his clients only. Because of this, we made an unwritten pact. We would keep our research separate. We collaborated only on activities that did not involve secrets or new research.

In spite of potential conflicts, we were able to negotiate with the University of Miami. They needed our rent money to help maintain the facility and pay the caretaker's salary. For fifty thousand dollars a year (it was called a service contract to avoid certain government regulations) we obtained one of the Spanish-style homes and turned each of the four bedrooms into an office. The building had been built of poured cement in 1929 and had a large living room with fireplace (Barbara's office), kitchen with working refrigerator and electric stove/oven, and enclosed front and back porches. We bought a "sediment-drying device" (microwave) to complete the kitchen appliances. What was really special was that the building was already wired into the FTS (Federal Telephone System). In addition, there was the modern five-room microbiology lab in the same building as the machine shop up the hill in the backyard. The hill was about four feet higher than the rest of the island. We set up a photographic darkroom, thin-sectioning equipment, X-radiographic equipment, and eventually a hydraulic press in that space.

We were off and running with the understanding that we would keep our little empire separate from Ginsburg's larger empire. We all rode the same ferryboat each morning and evening. It was a good time to plan the day ahead or talk about what we had done that day. At noon we all gathered at a picnic table under the sapodilla tree for lunch and watched the container ships and smaller vessels enter the Port of Miami through Government Cut. Lunchtime was when we discussed many things and solved the world's problems. We called our discussions "Guns and butter." It was a phrase we adapted because Ginsburg and his University of Miami grad students tended to be more liberal than us feds. We were unusual. Most federal employees lean to the left, but many in the Oil and Gas Branch, including Pete, had come from industry and saw things differently. "Guns and butter" was a phrase we would have much fun with over the years, especially when political issues came up. Everything was fair game under the sapodilla tree. The only subject not discussed was exactly what research the two elephants were pursuing.

Austrian-born geologist Wolfgang Schlager spent several years with Ginsburg on Fisher Island. He used to joke that Bob was the "King of the Island" and I was the "Viceroy." With Bob and his students there, many relationships and connections developed that would be long lasting.

The hourly ferryboat service was a double-edged sword. It kept walk-in visitors literally at bay, I should say across the bay! At the same time, it forced us to use our time most efficiently. It kept us moving when we did off-island errands. If you missed the boat by thirty seconds, you might sit on the dock in the hot sun for an hour until the next run. Bob Staysa, the caretaker/boat operator, would, however, turn around if you pulled up just as he was leaving the dock. After that, you were on your own. To solve that problem, we chained a small aluminum boat to the dock. If the current wasn't too strong, one could row across in ten minutes.

At Pete's suggestion, we placed a small USGS sign on the gate of the chain-link fence that surrounded our South Beach parking area. We agreed to let it be overgrown by vines. Our parking area and dock were on the south tip of Miami Beach. South Beach at the time was then a run-down crime-ridden area, but I knew it from those days of playing in small bands while in college. Even during high school, I had taken the bus to South Beach and spearfished along the Government Cut jetties. I would sell my catch to the Sanitary Fish Market on South Beach and purchase candy bars for lunch. Sometimes I slept on the beach to catch the high tide early in the morning. That's when the snook were abundant and the water was clear.

For privacy, we kept our office phone number unlisted, although some did find us to ask about sinkholes, earth tremors, and hurricanes. Some agency friends likened us to monks on an island retreat. It was great! We could focus entirely on our work. It was Heaven on Earth, or as Frank Lozo, a former Shell University geologist from Texas, used to say, "A bird nest on the ground." Who could ask for anything more?

Our hideout on the island functioned as we had hoped! During the first ten years we published sixty-five papers in peer-reviewed journals. We were very proud of that. Such productivity was not happening elsewhere in the USGS.

While Ginsburg's group focused mainly on the geology of the Bahama Bank and on teaching, we focused on the Florida Keys. Harold and I had practically grown up in the Keys and knew every rock and head coral, but there was something we needed badly. Our first order of business was to develop a diver-operated underwater coring device. No one had ever looked inside a coral reef in Florida without the use of dynamite or a bulky truck-and barge-mounted drill rig. Those kinds of rigs were very expensive.

Walter Adey and Ian MacIntyre at the Smithsonian had recently developed a hydraulically driven coring device. A large pump and a commercially available hydraulic wrench were the key ingredients. The system was connected together with flexible high-pressure hydraulic hose, and the drill pipe was commercially available. To learn more, Harold and I made a trip to Panama to watch Ian and Peter Glynn operate the device. At the time, Peter was with the Smithsonian field station in Panama. We were impressed. Theirs worked so well that we adopted the method but developed a smaller power source and, like theirs, used commercial drilling pipe (called rods) and core bits. The whole kit and caboodle could be transported and operated from our twenty-five-foot boat.

We named the boat *Halimeda*, after the calcareous alga that produces much of the sediment and limestone in the Florida Keys. We could trailer the boat everywhere or work directly from our boathouse. Over the years we took cores on reefs in Florida, Puerto Rico, Belize, the Bahamas, Bermuda, Dry Tortugas, the Philippines, and even in atomic bomb craters on a Pacific atoll. The bomb craters were at Enewetak Atoll in the Marshall Islands. The drill was equally useful above water, so eventually we were coring on land in Florida as well as on ancient reefs on mountaintops in New Mexico.

Our drill really got around. Had the drilling been contracted with drilling companies, costs would have been astronomical and the work simply would not have been done. What we accomplished for peanuts would have cost taxpayers millions—assuming we could have raised the money. Harold and I actually received an incentive award for development of the drill and for saving taxpayers money. Harold did much of the development. He found some heavy aluminum pipe and constructed a twelve-foot-high tripod to suspend the drill pipe and hydraulic motor. The three legs were held in place at the top by a specially made beryllium/bronze and stainless steel hinge. I had rescued that tripod head in the garage at the Shell Coral Gables lab years earlier. High-priced machinists at Shell University in Houston had carefully machined it—they spared no expense to make it. Unfortunately, it was overbuilt and way too heavy for our push-coring tripod, and besides, I had developed a single-legged monopod for sediment coring. We no longer used a tripod for sediment coring. Ginsburg had no use for it, so I confiscated it for our drilling tripod. That same tripod and drill are still in use at the St. Petersburg USGS office some thirty-five years later! So is the *Halimeda* that we had built to our specifications. As of this writing, the *Halimeda* has gone through two different inboard/outboard units and was

Gene and Dan Robbin coring the reef in Belize.

converted to a large outboard when they became available, and the boat has outlived five different outboards and is going strong.

In 1976 we drilled at Dry Tortugas on a National Park Service–funded project. While there we met a young fellow who was doing temporary work

Harold Hudson and Bob Halley coring the reef at Enewetak, with *Halimeda* above. This drilling rig went everywhere.

for a Park Service lobster researcher. He was a strong diver with a most likable personality. His supervisor, Gary Davis, wanted to see him land a permanent job, but at that time it couldn't be arranged within the Park Service. Luckily for us, we were able to hire Dan Robbin as our permanent geological technician. I started my career as a tech, so it was easy to relate to the job and to those who did such work. We had all shared technical work, but now we had a real technician. Our basic team of five was complete.

The day Dan officially joined the Skunk Works, we were drilling a reef off the lower Florida Keys from a chartered vessel. It was rough that day, but Dan caught on very quickly and thereafter we referred to that experience as his "trial by water." He passed with flying colors. Dan turned out to be a particularly good lecturer and originator of ideas, and we treated him like another researcher. He took classes at the University of Miami and became proficient with carbon-14 age dating.

Scorpion and Muleshoe Mounds

The drill created many adventures—it seemed to be at the root of them all—even the atomic bomb craters. An unusual adventure came far from Fisher Island—several time zones away in the mountains of New Mexico.

During my earlier Shell training and while living in Abilene and Midland, the Shinn family made many road trips. We had frequented the Sacramento Mountains in New Mexico, where I was impressed with the spectacular reefs and mud-mound buildups often visible from the highway. The reefs stood out against the mainly horizontal layers of limestone that formed the cactus-infested mountains. Many weekends were spent scrambling over these enigmatic features trying to avoid thorns. The ancient reefs and buildups were fascinating, but there is a limit to what the naked eye and a rock hammer can reveal.

James Lee Wilson had examined and published Shell Company reports on their composition, but the mounds had never been core-drilled, although it had been attempted. Some Shell University researchers tried coring with a gasoline-powered drill in the 1950s. They used a device similar to a noisy gasoline chain saw that turned a small core barrel. Success escaped them. The big hurdle was transporting water to the drill site, which was necessary for drilling. Air temperature on these reefs often exceeded 100°F, and it felt even hotter! The initial coring failure had become a legend within the company. The story so poisoned the whole idea that no one in the company ever suggested trying again.

What was different for us was our efficient hydraulic core drill. We had become very good at using it. The question was, could we accomplish what others had failed to do? What a grand challenge it would be! After all, we were in the Oil and Gas Branch, and there was a great deal of interest in these reef-like features among petroleum geologists. Reefs of similar age buried below ground only a few miles away were already producing oil! What had built the reefs? How had they formed? How could you tell where to drill to find the subsurface hydrocarbon-producing areas? We had some ideas to test. And even if our ideas were wrong, drilling would likely provide new information.

As a federal government agency, we were able to obtain a surplus Army trailer at no cost. With all drilling equipment loaded aboard, we headed west. By this time, Pete Rose had left the USGS and Bob Halley had transferred to the Denver office. So it was Dan Robbin, Harold Hudson, Barbara Lidz, and me. We were headed west on a most adventurous road trip. A young news photographer named Steve Earley would join us later. Our first stop was in the Texas hill country. We would visit Pete at his family ranch in Telegraph, Texas, where we would spend the night and enjoy a great Texas barbecue. We had a lot of catching up to do and many stories to tell!

Bright and early the next morning we continued on to Alamogordo, New Mexico, where we checked into the Sands Motel. It was a favorite motel for visiting geologists. The next day we drove north to Tularosa, New Mexico, to examine and work out logistics. Were we surprised, or should I say I was stunned and disheartened, when we arrived. The outcrop of interest was farther up the hill than I had remembered. We would have to get the equipment, including water and drill mud, up a steep outcrop to reach the drill site. The target was one of the famous Permian reefs that had been studied by many using hammers and hand lenses. Each reef mound was approximately fifty feet thick, but they were more than three hundred feet up a steep cactus-covered slope. We were appalled at the daunting task. How in the world were we going to lug our equipment up there?

And there was that old bugaboo, water. We had two fifty-five-gallon drums of water we needed for drilling. Water was absolutely essential. The task looked as though it might take a week just to get the stuff up there! Dejected, we returned to the motel where we moaned and groaned, to say the least.

On a hunch, we looked in the yellow pages under "helicopters." There was one listed, but it wasn't in Alamogordo. We called the number anyway. A very pleasant guy answered. Yes, he could fly down to us, and yes, he could probably transport our equipment up the hill. It was the kind of thing he had done in Vietnam. But how would we pay? Ah, we had a secret weapon! We had with us a book of Form 44s! They were a special kind of purchase order for emergency government use. The pilot knew all about them, as he had done work for the Bureau of Indian Affairs. What we didn't know was that we were breaking a whole drawer full of USGS rules and regulations about helicopter use. First, we were bypassing the aviation department that handles air and helicopter contracts. We didn't even know of its existence then. That alone, as I would learn later, takes several weeks, assuming the equipment had passed inspection and was approved for government use. We had no time for that. Didn't I say our motto was, "'Tis easier to seek forgiveness than to seek permission"? We had come all this way and were not stopping now!

We cut a deal, and the next morning we met at the base of the hill. By noon, all of our equipment—including the two drums of water—was on the drill site, the apex of what geologists know as Scorpion Mound. We even had heavy bags of drill-mud mix. By 5 P.M. we had the first five-foot length of core in the box! For the next two days we simply hiked up the hill (it took a little less than an hour) and began drilling by 7:30 A.M. I remember one day very well. It was Halloween, and we wore rubber masks while drilling to celebrate the day!

At a nearby department store, we had purchased a four-foot-diameter plastic kiddie pool and some cactus-proof boots. The pool cost about four dollars. In the foot-deep pool we mixed drill-mud powder and water. The thick slurry was circulated down the hole with a small gasoline-powered pump. We felt just like real oil drillers. Soon Steve Earley, the news photographer working in Odessa, Texas, joined us. Steve, who had been with us before, had been a diver/underwater photographer provided to us by an organization in Chicago called "Our World Underwater." He had won their scholarship earlier and had traveled with us for an underwater-drilling expedition in the Philippines. He was free! All we had to do was supply 35-millimeter film.

What did we learn, and why did we choose this place? These reef mounds near Tularosa had been examined by hundreds of geologists over the years as part of their training. The reef mounds had accumulated during the Permian, a time of great reef building in Earth history. A lot of oil had been pumped from Permian reefs all over the world. However, these reefs were different. Calcified algae about the shape and size of potato chips had

Drilling a Permian algal reef, locally known as Scorpion Mound, in New Mexico. Barbara Lidz (*foreground*) examines and describes cores as they come out of the ground. Photo by Steve Earley.

constructed these reefs. The calcified chips seemed to float in a lime-mud matrix and, most curious, were riddled with fractures.

Years earlier, Bob Dunham had termed such rocks "collapse breccia." He tied their origin to his thesis that freshwater dissolved holes in the rock. The

Field geology in New Mexico. *Left to right*: Dan Robbin, Gene Shinn, Barbara Lidz. Photo by Steve Earley.

holes made the rock unstable, and because of overburden the rock collapsed in on itself. In some reefs those holes held oil. Dunham had published a famous paper on this kind of rock. The collapse breccia he described was in an oil-producing feature known as Townsend Mound located in New Mexico. So here we could see the same features in our core, and we could also walk along the outcrop and trace the layers for more than a mile. We could also see a curious layer about a foot thick that underlay this unit.

One of the first things we discovered while coring was that the layer beneath the mound was about five feet thick. It was only a foot thick at the outcrop. Without the core, no one could know it was significantly thicker beneath the mound. We had drilled just twenty feet back from the outcrop. So right away we deduced that the reef had grown there because of a preexisting topographic feature. Everything was consistent with what Bob Ginsburg had recently discovered under much younger coral reefs in Belize. The topography underneath was also consistent with what we had been finding in the Florida Keys. Reefs in the Keys are located over preexisting topography.

It was also apparent that the Tularosa reef had been converted to limestone while it was still growing. That was new. Percolating freshwater had not created the so-called collapse breccias. In fact, the rock was not collapse breccia at all! It was a type we would later reproduce with a hydraulic press. We would simulate the rock by compressing a mixture of mud, portland cement, and broken chicken eggshells. We were happy campers with our new discovery!

Next on our list was a famous mound to the south called Muleshoe. Of Mississippian age—much older than the Permian mounds—Muleshoe reef forms a three-hundred-foot-high knob on a prominent mountain south of Alamogordo. It was in ranch country, and what a bumpy, axle-breaking journey it was just to reach the base of the reef. Again the helicopter met us and flew the equipment to the distant reef top. This site was much more remote, and each morning it took us an hour to hike up the flank of the mound. The cores from Muleshoe were even more spectacular than those from Scorpion. The rock was so dense that the five-foot-long cores (that's the length of a core barrel) had to be broken to fit them in the short core boxes.

By the time we reached a depth of seventy feet, the diamonds on the core bit finally gave up—they were worn off. We could go no farther, but we were pleased. We had about reached the limit of our drill unit anyway. What we found was spectacular! The cores presented an entirely different picture from that seen at the outcrop. The outcrop near the base of the mound is vertical and looks like featureless lime mudstone. A reasonable-size sample is virtually impossible to break from the vertical wall. Many scientific papers have been published about this "mud mound," and most describe it as lithified lime mud. Some compared it with the mud banks in Florida Bay. But, mud was not what we found in our cores! We found delicate lacey bryozoa in near-vertical growth position. The organisms seemed to be floating in crystalline calcite. It definitely was not mud. Imagine suspending a string (a bryozoan) in a concentrated sugar solution and letting sugar crystals grow on the string. Then let the sugar continue to crystallize until it all becomes one hardened mass. It seemed that calcite had precipitated on the bryozoa in the ancient Mississippian sea just as sugar precipitates from a supersaturated solution.

We had expected to find crinoids—the distant relatives of starfish which look like flowers on tall, segmented stems. During the Mississippian, their stems generally broke into barrel-shaped fragments. There were a few crinoids in the cores, but not many. Apparently, as the mound was growing, it became cemented into rock (think of the sugar-and-string analogy), and the crinoids that grew on its surface were periodically broken and washed off the reef crest. Their fragments formed lime sand that had accumulated in the low valleys around the reef. The crinoid sand around Muleshoe Mound is many meters thick. We concluded that the mounds themselves would make poor oil reservoirs but the flanking crinoid sands might serve as reservoirs if the oil entered early in the sands history. Our conclusion was, if found in the subsurface, do not drill the center but instead the flanks where the crinoid sands would be located. The cores are stored in the USGS core library in Denver and have been examined by many geologists.

Upon return to Fisher Island Station from the New Mexico adventure and revealing research, I had my hand slapped, as expected, for using an unauthorized helicopter. However, it was difficult to reprimand us in the light of such success. Did I mention that this project had been very inexpensive? We had done an excellent job, no one was hurt, all of us had all of our fingers and toes, and we had cores in the box. Maybe there were a few cacti punctures. The bill for the helicopter was fourteen hundred dollars.

Captain Roy, the *Sea Angel*, and the *Captain's Lady*

A major key to our success at Fisher Island was our association with a diver, boat captain, salvager, former treasure hunter, towboat operator, and master mechanic named Roy Gaensslen. We called him Captain Roy. It was aboard his boat that Dan Robbin survived his trial by water. Roy's first boat was a fifty-foot, single-engine, wooden, converted shrimp trawler called *Sea Angel*. Being a government agency, we always had to solicit bids on boats we chartered. We did it many times, but no one could ever touch Captain Roy's low bid. We generally conducted our drilling and diving from the USGS boat *Halimeda* but used Roy's boat to carry cargo, coring supplies, scuba tanks, compressors, fifty-five-gallon water barrels, you name it, and to tow the *Halimeda*.

More than anything, *Sea Angel* served as a floating dormitory and on-site restaurant. Not air-conditioned, she could sleep six if the below-deck forward bunks were used. We spent many nights anchored at the work site when weather permitted, saving time and money that might otherwise have been squandered on hotel rooms. The real advantage was that there were no excuses for not getting in the water early each morning. On top of all that, Roy could fix anything that broke. The tools he could fetch from the bowels of the *Sea Angel* were amazing. He even had welding and cutting torches!

On Roy's boat we made two different seven-hundred-mile voyages to Belize in Central America. These were trips that lasted a month. In Belize we took sediment cores and core-drilled the largest barrier reef in the Atlantic. The boat cost the princely sum of $125 per day! No one else had a boat for less than three times that much.

Eventually, Roy moved up to a fifty-two-foot, twin-engine, three-stateroom, air-conditioned Marine Trader. Barbara Lidz helped him name the vessel *Captain's Lady*. Roy was literally like part of our family. With our budget, we could never have accomplished what we did without Roy's help, patience, expertise, and eager scientific interest.

Roy succumbed to pancreatic cancer in 1997. Thanks to Barbara and Harold's dogged determination, there is now a reef in the Florida Keys National Marine Sanctuary named in his memory in honor of his many years of having contributed to Florida Keys science. Harold constructed an underwater memorial, and we spread Roy's ashes there a few years later. It was a somber, heartrending ceremony being conducted from my new trawler, *Papa-San*. Roy had helped me choose the trawler. After his death it was then equipped with the stabilizers, called "flopper stoppers," that had come from the *Captain's Lady*.

One of my purposes in writing this book is to reveal the inner workings of geological science, the conflicts and controversies—and how science really works, and sometimes does not—and how much fun it can be. Oh, what adventures we had! But overall, I want to stress the importance of having incredible colleagues and gifted friends like Captain Roy to draw upon. This is the part about science that is seldom told in science biographies.

Whitings, One More Time

While we were at Fisher Island, Paul Enos, Bob Halley, Randy Steinen from the University of Connecticut at Storrs, and I began conducting weeklong field courses for the American Association of Petroleum Geologists. The last three days of each course were spent in the Bahamas living aboard the *Sea Angel*, and later on the *Captain's Lady*, with Barbara as chief cook and bottle washer on both vessels. On these trips we always passed over a sandy part of the Great Bahama Bank en route to the central muddy part of the bank. Our destination was the humid horsefly- and sand fly–ridden tidal flats of western Andros Island, and we would go ashore using rubber boats.

In the early 1980s we began noticing more whitings—the patches of milky water surrounded by clear water mentioned earlier—in places where they had previously been uncommon. We knew the sandy bottom could not be stirred into suspension by fish or anything else. Farther out on the bank where the bottom was muddy was different; there, it was easy to stir up the mud. Many geochemists, people who study the chemistry of rock and water, had for the most part agreed that calcium carbonate could not precipitate directly from seawater. Some would say the mud had been stirred up from the muddy bottom and had drifted over the sand. Simple experiments disproved that idea.

We became more and more curious. In 1983 we had some leftover funds. In government service, these funds were called "end-of-the-year money." You use it or lose it. With this money, we organized an expedition to the Bahama Bank aboard Captain Roy's boat. Roy's son Jim, an accomplished pilot, flew ahead in his private plane and served as a whiting spotter. He saved us time by directing us to the best whitings.

On this whiting trip to the Bahamas, our objective was simple. Rather than do geochemistry, we would simply look for the fish that geochemical "experts" said had to be there. To this end, we had borrowed a side-scan sonar unit from another USGS office. We had also borrowed a shrimper's try net from National Marine Fisheries. A try net is a small version of a regular shrimp trawl and is used to see if there are enough shrimp to justify putting out the big nets. Our plan was to make timed hauls in the clear water and then drag it in the muddy water for the same length of time. If there were fish, we should catch them. We reasoned that fish in clear water could see the net coming and get out of the way. In muddy water they couldn't see the net.

The side-scan sonar was used to see through the muddy water and to spot schools of fish. We never found any, with either the net or the sonar. In addition to the search for fish, Dan Robbin had constructed an underwater flume made of clear Plexiglas that we mounted upside down on the bottom.

Think of a square tube about ten feet long that is open at each end and along one side. The open side is pegged to the bottom. Inside we placed an electric, variable-speed, underwater scooter and a current meter. We could increase the current flow over the bottom and measure how fast a current was needed to begin moving the mud. We soon determined that normal tidal flow was not nearly enough to lift mud from the bottom.

We had some real excitement on one occasion. A shark bit and bent the propeller on the current meter we had deployed in a whiting. Bob Halley's eyes were big as saucers when he emerged from the milky water with the bent meter in his hand! In some whitings, he had sat on the bottom for an hour to watch white precipitates fall from the water column onto his black wet suit. The bent-propeller incident made us leery of diving in whitings thereafter.

Dragging the try net indeed caught more fish in the whitings, but not the kind that feed on the bottom. In the clear water, any fish present simply zipped out of the way, and these fish had been caught simply because they couldn't see the net.

By dragging the net, with its heavy chain and otter boards, in a circle we could make our own whitings. We learned a valuable lesson doing that. Artificially made whitings settled to the bottom within four hours. Natural whitings nearby that we could watch at the same time never settled for as long as we watched! If we pumped natural whiting water into large containers aboard the boat, the sediment would settle out in about four hours! This meant that a whiting could not drift very far before it settled to the bottom. They must be continually forming near the surface. These were simple observations that drove us to do more tests. We even used fish poison—still no fish!

Side-scan sonar did not detect fish, and neither did off-the-shelf fish finders. However, we did discover that large whitings contained numerous inquisitive sharks. They were snow-white black tips and practically invisible within the whitings. That's what had attacked our current meter. Small whitings did not contain sharks, so we felt safer swimming in them. When we dove to the bottom and let the muddy clouds pass over, we would be covered in mud. Mud was continually raining down—not going up! It became obvious that mud was being continually generated near the surface. Whitings somehow regenerated themselves as they drifted along, depositing sediment like a rain cloud drizzling mist or water droplets.

When there was no end-of-the-year money, we switched to bootlegging trips. We were thoroughly hooked on the mystery, even though whitings did not seem important or scientifically relevant at the time. Time would prove otherwise. To keep the study going we went underground, so to speak. One way was to get Sonny Gruber hooked! Gruber was and still is South Florida's shark expert and is located at the University of Miami Rosenstiel School. He had studied sharks in the Bimini lagoon for years but was unaware of the white black-tip sharks in whitings. In fact, he thought we were crazy until we brought him a small one in an ice chest, which Jerry Koch, another graduate of Shell University, and I had caught. Once interested, Gruber obtained a long "gill net" used for netting mullet. Mullet try to swim through the mesh and become caught by their gills. They cannot back out once trapped.

Randy Steinen and I went to the Bahamas on a University of Miami vessel, and Sonny brought along an ex-mullet fisherman. The twenty-foot-wide net we used had corks along the top and lead weights along the bottom. We would watch to determine which way a whiting was drifting, then would stretch the net in front of the drifting whiting and let the muddy water pass through. The only fish we caught was one baby shark. Meanwhile, the whiting continued drifting along unabated. If there had been a school of fish making that whiting, wouldn't they have been caught in the net? Wouldn't the mud have settled to the bottom? That seemed logical! Or was it overly simplistic? That's what a lot of armchair scientists thought!

In another experiment we dispensed rotenone, a powerful fish poison. Fisheries biologists have used rotenone for many years. To dispense it, we dragged a clear plastic hose with a five-foot-long length of perforated iron pipe at the bottom end. The perforated pipe bounced along the bottom as the poison was dispensed through the perforations. Any mud-stirring bottom fish would receive a deadly dose. What happened? The whiting just drifted along unabated. A few small fish were killed, but they were not bottom-feeders.

After all this, we were beginning to understand why no commercial fishermen had been attracted to these places. Fishermen do catch mullet in whitings in Florida Bay, but there are no mullet in the Bahamas. One thing you can rely on, if commercial fishermen had known there was something worth catching in Bahamian whitings, they would have been there long ago. Nevertheless, all the Bahamian cruising guides still say, "The patches of milky water are caused by schools of fish!"

A logical, nonscientist fisherman would appreciate the measures we used to verify the lack of fish. Unfortunately, that is not always how science works. Dragging nets and using rotenone are not what normally come under the heading of science—we were not wearing white lab coats! And besides, real scientists had worked on whitings before. They had made chemical measurements of alkalinity or had examined bottom sediment. To them, what we had done would simply be called making anecdotal observations. But didn't Darwin come up with the theory of evolution using anecdotal observations? The reader, unaware of the long-standing arguments over whitings, wouldn't know we were going against the conclusions of dozens of famous geochemists who had come before us. They had long ago concluded that whitings could not be caused by direct precipitation of calcium carbonate. That's what had been taught in classrooms for several years. For that reason, we went after the fish hypothesis as a way to force geochemical experts to look the phenomenon straight in the eye.

I gave many talks on whitings, as did Randy Steinen. They were popular talks but didn't change any geochemical theories. College students, however, loved the controversial story. I was learning another good lesson on how science works—or doesn't.

Drug-Smuggler Encounters

Trips to the Bahama Bank were always interesting. Roy's first boat, *Sea Angel*, resembled a typical "mother ship." During the drug-dealing days of the 1970s and 1980s, boats that brought bales of marijuana or cocaine from Central America were called "mother ships." They would head for the Florida Keys or the Bahamas, where they would meet up with smaller, faster boats. At the same time, small airplanes would off-load their product at secluded airstrips. Because of the mother-ship appearance of the shrimp trawler, *Sea Angel* was often stopped and boarded by the Coast Guard or the Drug Enforcement Agency (DEA). They always called the boarding a "safety inspection."

We had nothing to worry about, but these checks sometimes wasted a lot of our time. It was no fun being ordered against the rail by a hyperactive young agent with an automatic weapon! DEA agents didn't always wear uniforms, and their boats were generally not marked. They looked just like real smugglers. Coast Guard personnel were always in uniform and wore highly polished black shoes.

The really scary times occurred when fast "cigarette" boats, called "go-fast" boats, approached us. Were they the real drug dealers, or were they DEA? Either way, they thought we were a mother ship. Generally, they were easy to recognize when they got close. Real smugglers wore heavy gold chains. DEA agents did not! When they came close, my job was to load the AR-15 and keep it hidden while Roy went on deck to convince them we were not the boat they were looking for.

We saw airdrops in which the bales of marijuana would rain from the sky, making splash after splash. I once videotaped an aerial encounter between a Black Hawk helicopter and a two-engine drug plane. Helicopters were known to squat over a plane and force it down into the water. This one got away. The Bahama Banks remain littered with submerged and semisubmerged drug planes. I once wrote a one-page story about an encounter for a Geological Society of America series called "Geotales."

Yet Another Whiting Story, Not the Last!

While we were investigating whitings we attracted another researcher with a new idea. She was Lisa Robbins, a graduate student doing her dissertation at the University of Miami under Cesare Emiliani. She had used proteins to identify foraminifera by analyzing proteins and amino acids within their dead calcium carbonate skeletons. Lisa thought the technique might be a novel way to determine whiting origins.

If the mud originated as the aragonite needles within plants, such as *Penicillus*, then the calcium carbonate in them should contain the same proteins or amino acids as in the mud-making plants. Robbins was awarded a postdoctoral scholarship to work with us on the problem. It didn't take long to determine that plants did not make the aragonite needles found in whitings. Nevertheless, her discoveries did not affect the way conventional geochemists thought. Organic chemistry applied to calcium precipitation was too novel for them to accept.

Lisa later wrote several proposals to the National Science Foundation, but each time the ruling geochemists shot them down. "Everybody knows they are stirred up by fish, so why waste research dollars on them?" was a common remark. According to an informant, she once got good reviews but the person in charge decided she was not the right person to do the work! It was frustrating to say the least!

In the meantime, we finally managed to publish a peer-reviewed article titled "Whitings, a Sedimentological Dilemma." It was published in the *Journal of Sedimentary Petrology* under the heading of a "perspectives paper." We called it a dilemma because that was the only way to get it published. I vividly recall some sixty-five different written questions from one of my best friends. All had to be addressed to get the paper accepted for publication. I realized that even though we were only disproving the long-standing fish theory, the major implication of our observations was that whitings had to originate in the water column. If that were accepted, then the onus would be on the geochemists to explain the phenomenon.

The paper was coauthored by Peter Swart, a renowned geochemist at the University of Miami; Randy Steinen, a geologist from the University of Connecticut; and Barbara Lidz, my close USGS colleague. That paper quickly caught the attention of a potential funding source. It was not a source one would expect—the Electric Power Research Institute (EPRI) in California!

Others who had aided the study in the beginning, including Bob Halley, were beginning their careers and were leery of involvement in such a controversial subject. They dropped out early. This was another lesson about how the enterprise of science works. Young, emerging scientists often shy away from potentially career-limiting subjects, and they can't be faulted for that.

Was it really that controversial? Randy Steinen had submitted three separate proposals to NSF over several years and was rejected by reviewers each time, as Lisa Robbins had been. Randy's approach was to determine if there were sufficient plants on the Bahama Bank to explain the amount of mud that was there. The truth was that we had already proved there were not enough during bootleg trips. His approach was considered anecdotal—not real science. Randy also wanted to show that strontium content of whiting mud was different from that in aragonite made by plants. We already knew the answer. Lisa had previously concluded the amino acids in mud-producing plants were different from those in the whiting mud. Why was all this information not accepted?

The problem waited for observations made from a research submersible in deep off-bank waters some years later. These observations showed how wrong all those reviewers had been. In the meantime, Lisa had teamed up with Pat Blackwelder at the University of Miami. Using transmission electron microscopy, they discovered that clumps of aragonite needles were clustered around algal cells. They were cyanobacteria!

After the perspectives paper was finally published, I received a call from EPRI. The fellow said they wanted to fund more work on whitings! We had just moved our office from Fisher Island to the new St. Petersburg USGS office and were engaged in other research. Besides, I saw no way their organization could fund a government agency. Just what was EPRI's interest in the first place?

The fellow explained that if precipitation from the water was occurring, it might be a sink for carbon dioxide, and the industry might use whitings as mitigation against power-plant emissions. Wow! They were thinking way ahead and along much different lines than anyone else. It was a good example of the forward thinking that private-research labs generally do. It reminded me of the way we had worked at Shell University.

Lisa Robbins, by then, was a newly graduated young professor at the University of South Florida in nearby Tampa. I suggested they call her. If she was right about cyanobacteria being the cause of the precipitation, then the organisms might be manipulated to create even more carbon dioxide drawdown. They called her.

A week later, Lisa made a presentation in California and almost overnight received funding for her work. It wasn't long after that she was practically supporting the entire geology department. Before it was over, she had brought in a little over a million dollars for her department. One of her students, Kim Yates, did her dissertation on how cyanobacteria blooms stimulate precipitation. Having a biological component partially satisfied the geochemists, but not entirely. There were still holdouts, such as Wally Broecker at Lamont and John Morse at Texas A&M University.

Another Approach

Much had happened since our 1989 move to St. Petersburg. I would return to the whiting controversy once again but not until the twenty-first century, mainly because I had to wait until I acquired my own live-aboard boat. The forty-two-foot trawler would allow independent travel to the Bahamas, but more important, the study also needed the help of a geochemist. I found one. His name was Charles Holmes.

Chuck, a fellow USGS scientist, had pioneered the use of a natural short-lived isotope of beryllium called beryllium 7 (Be-7). Be-7 emits gamma rays that can be detected using a gamma counter. After fifty-three days, the gamma radiation from Be-7 is reduced to half. In other words, it has a half-life of fifty-three days. Chuck had used this isotope for sedimentation studies. How did it work? If sediment contains abundant Be-7, then it had been deposited more recently than sediment that had been lying on the bottom for more than fifty-three days. Chuck recognized a possible application for solving the whiting controversy using Be-7.

His idea was straightforward. Lime mud collected from a whiting precipitating from the water column should contain more Be-7 than mud stirred from the bottom where it had lain for several months or years. Be-7 is continually falling naturally from the sky, as it were, and its content in seawater is constantly being replenished and thus remains relatively stable. It would therefore be incorporated in any minerals precipitating out of that water. Bottom sediment should contain less because of its short half-life. If stirred to make a whiting, that sediment should not contain nearly as much Be-7 as that precipitating from the water.

So off I went to the Bahamas with large filters through which I could pump seawater from whitings. I also sampled sediment from the bottom. In addition, I made a device with chains similar to the try net we had used to make artificial whitings earlier. I could drag this device and steer the boat in tight circles to continually stir up bottom sediment. The artificial whiting mud was pumped aboard through the large filters as the boat circled within the cloud of mud. What did we find? The result of three separate self-funded expeditions was this: sediment from the natural whitings always contained more Be-7 than bottom sediment, including that stirred into suspension with my "artificial fish-school device."

The filtered samples were always counted for gamma emissions within three weeks of collection. Again, these were "bootlegged" analyses not supported by external grants. Chuck's technician, Marci Marot, operated the equipment and conducted the analyses. The data were presented at a national meeting, where there were mixed reactions to the new information. Some of those in the audience thought I was beating a dead horse—they already knew whitings were precipitated—others remained unconvinced. There were also those who thought the issue interesting but of no practical significance. Another criticism was that there should be more mud on the Bahama Banks if precipitation was the mud's origin. The thickness of bank-accumulated mud is only about two meters! There should be more! Why wasn't there more?

That criticism came mainly from laboratory geochemists unaware of what happens when a storm or a hurricane passes over the Bahama Banks. Hurricane Andrew in 1992 swept mud off the bank in huge amounts. The fact that mud was swept off the bank was proven by published studies conducted using seismic profiles, sediment coring, and actual observation with the research submersible *Delta*. There is more mud in deep water at the base of the Bahama Bank margin than on the entire Bahama Bank. Mud swept off the bank has formed a wedge more than a hundred feet thick beginning at about seven hundred feet below sea level. But this was nothing compared to what would eventually be found there using a deep-diving manned submersible, the Johnson *Sea-Link*, operated by the Harbor Branch Foundation. Yes, that's the same submarine that had been stuck on the shipwreck near Key West many years earlier. I continued to study whitings even though no one saw any usefulness for this academic study, and besides, the implications were still a problem for conventional geochemistry. If the whitings are really precipitating from within the water column, many basic tenets of carbonate chemistry were being violated.

And then along came Christopher Kendall, the same fellow who had been working on the sabkhas in Abu Dhabi while I was working in Qatar back in the mid-1960s. Chris knows the geology of the Middle East oil province better than I. However, we both knew there had long been a problem explaining the source of Middle East oil. The known sources for the oil were carbonate rocks deposited during Permian and Cretaceous times. Just the possibility of carbonate source rock for petroleum in itself had long been controversial. What whitings might explain about Middle East oil made this a whole new ball game, as we saw later.

The Lighter Side

Fisher Island wasn't all work and no play. We watched the whole Art Deco movement play out on South Miami Beach. We drove our cars through it every day to catch the ferryboat over to the island. We also watched from our front porch across Government Cut. Often there were nudes sunbathing on the rock jetty across the channel from our office. We acquired a telescope for our front porch!

For the first few years before the Art Deco movement, South Beach became home for hundreds of so-called boat-lift people that Castro expelled from Cuba. The tobacco crop had failed due to a strange blight, and many workers were out of jobs.[2] Partially because of the blight, Castro allowed people to leave the island, and at the same time he cleaned out the jails and mental institutions. As a result of the sudden influx of "boat-lift" people accumulated in the already run-down South Beach area, there were soon several murders taking place there every week. Some said it was the law-abiding anti-Castro Cubans taking care of the questionable types let loose in South Florida. After that period passed, the old hotels—the same ones where I had played in dance bands during my college years—were taken over and transformed by a new, mostly gay crowd. They adopted the colors and theme of the television series *Miami Vice*. Earth tones were out—white, pink, and chartreuse were in. Old hotels were transformed, seemingly overnight, into swishy places where the upwardly mobile beautiful people could be seen nightly. It was also a cocaine dealer's haven—or was it heaven? Some of the *Miami Vice* episodes were based on events taking place on South Beach. Apartment buildings from the 1930s were coming down in avalanches under the heavy impact of demolition balls. Watching all this, I decided it was a good time to write a humorous spoof on South Beach, which centered on the mayhem after roaches nearly became extinct.

I wrote the story for *Tropic* magazine, a *Miami Herald* Sunday edition insert that each year published a "cockroach edition." It was mainly devoted to politicians whom the editor, syndicated humorist Dave Barry, thought deserved the annual "cockroach award." As fate would have it, I wrote the story just as Dave Barry was changing jobs. *Tropic* couldn't use it. My next thought was that either the Shell or Exxon in-house magazines might publish the spoof. I still knew the editors of both. The Exxon magazine editor said, "Gene, we can't publish this! It's too close to the truth!" So I gave it to the Houston Geological Society, and they printed it in their October 1989 monthly bulletin.

Mashing Mud

One of the prevailing paradigms of the day, one that had had large implications during my Shell days, was that carbonate sediment, especially lime mud, does not compact under the pressure of overburden.[3]

The mud-bank-sediment cores that I had been impregnating with resin and slicing for observation back in my Shell days never showed signs of mashing or distortion. Burrows remained circular, and even unfilled open burrows remained open. Remember the plastic I poured down crustacean burrows? That alone indicated that the burrows in soft sediment had remained open and resisted distortion. Many sediment cores contained open burrows that had been overlain by several meters of sediment for three to four thousand years, yet the burrows had not been mashed flat! Because of these observations, I had no reason to doubt the ruling theory that mud does not compact. The problem was that when we look at ancient shallow-marine lime rocks, they often contain mashed burrows, wavy organic stringers, and the rocks are bedded.

We didn't see those features in the impregnated sediment cores. The impregnated sediment cores were a great step forward in understanding ancient limestone deposition, but something wasn't quite right.

The Fisher Island setting allowed for experimentation without seeking additional funding and without having to explain everything we were attempting. For example, the belief that lime sediments do not compress was so ingrained that I knew better than to ask to test the theory. What I did certainly would not have been attempted at Shell University. At the USGS, explaining why I wanted to experiment probably would have gone nowhere. Some initial proof of compaction was needed. So, how to get some proof?

While developing the underwater drilling device we made friends with a machinist named Bus Headberg, who had a well-outfitted machine shop. He also took a special interest in our work. He found our projects much more interesting than fixing automobile engines. In addition to being a good machinist, he was inexpensive and went out of his way to make useful design suggestions. In the center of his shop was a huge hydraulic press, and that gave me an idea. One day I asked Bus to machine some stainless-steel pistons with ports that would allow the escape of fluids. He quickly got the drift of what I wanted. He fitted the pistons with O-ring seals and machined them to fit in four-inch-diameter core tubes.

My plan was to take some short cores, about two feet in length, in environments where we had previously taken cores. After taking several cores on a mud bank in Florida Bay, we fitted the specially made porous pistons in both ends of the core tube. I then took these to the machine shop, and at no charge, Bus put them in his big press. I thought he would throw me out when the O-ring seal burst and sprayed mud all over the ceiling. He didn't mind, and we persisted. Sure enough, water was expelled through the relief ports as the pistons sank farther and farther into the mud-filled tubes. After exerting several thousand pounds of pressure, the water ceased to flow and the mud would mash no more.

Back at the Fisher Island laboratory, we sawed the tube in half. We had planned to impregnate the compressed core with plastic resin, but that wasn't needed. The sediment was so tightly compressed that we could slice it just as if it were a rock. Surprise, surprise! The compressed mud looked more like a typical ancient limestone than any of the impregnated cores we had ever seen! Vertically oriented sea-grass rhizomes and roots were no longer visible. They had been mashed and squeezed horizontally to form thin organic films and wisps. This was really exciting! We immediately took close-up photos and made thin sections.

Another surprise! Delicate fossils, mollusks, and foraminifera had not been mashed! The sediment had simply flowed around them. Why were fossils not squashed? Think about divers underwater. They are not wholly flattened by water pressure. The pressure is uniformly distributed. The mud had behaved like viscous water.

We prepared a short paper for publication in *Geology*, a new journal that had been launched to announce new and often controversial discoveries or ideas. The paper, "Limestone Compaction, an Enigma," was well received. Every geologist who saw the published photos instantly recognized that the compressed sediment looked like typical ancient limestone. I must say that at that time more geologists actually looked at rocks than they do today.

The paper gave us the ammunition we needed to obtain backing so we could purchase our own hydraulic press, which would be set up in our lab space on Fisher Island. With the hydraulic press, we could do additional controlled experiments. It also had the advantage of being inside an air-conditioned laboratory.

We also made a simple mechanical press from steel "I" beams. The press was too big to fit in the laboratory, so it sat outside in the sun and weather. At the end of a twenty-foot-long I-beam lever, we suspended a fifty-five-gallon drum to which we could add various amounts of water. With this

Rube Goldberg device we could put a sediment-filled core tube—with pistons—near the fulcrum point and vary the pressure by the amount of water added to the drum. The press would produce variable pressures simulating burial up to about eight thousand feet, and it maintained constant pressure. It was simple and entirely mechanical. Gravity did all the work.

Making Oil

Dan Robbin was literally turned loose on the compaction project, and he thought up many different ways to do the experiments. By then we had gained a lot of attention, and calls came in from geology friends in the petroleum industry several times a week. The theory that only shale compacts was being seriously questioned.

One discovery usually leads to another, and this was no exception. A geochemist within the USGS Oil and Gas Branch in the Denver office, Jim Palacas, had previously taken samples of lime mud from Florida Bay and had tested them for organics. He placed them in a sealed pipe-like device. The test is called a Fischer Assay. He heated the little sediment-filled bomb in an oven. By doing this, he actually created crude oil. Palacas calculated that each ton of Florida Bay mud could theoretically generate about three gallons of oil. For comparison, oil shale produces a lot more, around fifteen gallons per ton.

We of course knew that in nature buried sediment is exposed to increasing temperature and pressure with time. We would do an experiment that simply sped up the process. At the time we were not aware that additional pressure is created as organic material is being converted to oil, and that the pressure generated would help drive the oil from a source rock to a porous reservoir rock. Many petroleum geologists had believed lime mud could not compress, and this is one of the reasons so few people understood the origin of oil in the Middle East, where there was little evidence of shale source beds. We were indeed showing that some old ideas were wrong. We tried yet another experiment, the results of which were totally unexpected!

It seemed logical that if we put a heating coil around our device and heated the tube while at the same time applying pressure, we might convert the sediment to a real rock. We suspected that through heating we might actually convert the aragonite (the unstable form of limestone) to calcite, the stable form, and the recrystallization process might produce real lime-stone. Who knows? We might even make dolomite, which to this day remains a great geological mystery. This was the kind of experimentation—call it alchemy—we could do in a remote setting on an island away from curious eyes and higher management. No one told us it wouldn't work. No one knew! If I had been more of a geochemist, I probably wouldn't have tried. What happened? The experiment did not make a rock. It made something better!

When we removed the compressed core, it was covered with a dark tar-like substance. The stuff fluoresced under ultraviolet light—a sure test that we had produced hydrocarbon! With the help of Jim Palacas, a real hydro-carbon geochemist, we determined that we had almost made real crude oil, and the tarry substance was chemically almost identical to the hydrocarbons Jim had analyzed from Upper Cretaceous limestone in Florida. We had made something that was almost real oil in only one week! In nature it takes millions of years. I must confess it smelled more like burned paper than oil.

With George Claypool, another geochemist at the Denver lab, we published our results in a special publication—AAPG Studies in Geology No. 18, *Petroleum Geochemistry and Source Rock Potential of Carbonate Rocks*, and Jim Palacas edited the volume.

Meanwhile, Dan continued to experiment, conducting a series of compaction experiments that simulated various depths of burial. We eventually published that work in the *Journal of Sedimentary Petrology*. By then we had also found evidence of actual pressure dissolution, which is what happens when constant pressure between two grains causes them to dissolve into each other. It is common in ancient limestone and requires millions of years to happen. Remember the jawbreaker experiment performed by Bob Dunham that I described earlier?

We speculated that the wispy organic layers looked so much like stylo-lites that, given geologic time, they probably would evolve into true stylo-lites.[4] During my Shell days I had examined many stylolites in outcrops and in cores from oil wells, and I knew they could be associated with hydrocarbons. In some areas stylolites often showed evidence that the process had caused complete dissolution and disappearance of up to a foot of limestone. Might this be a mechanism for expelling oil from fine-grained limestone?

Any oil in the limestone that dissolved had to go somewhere. In our publication we speculated that was the case.

Squeezing mud, as we called it, was something we did during the early 1980s when the oil industry was riding high and there was much interest in everything else we did at Fisher Island. It was a unique time, and best of all we could call our own shots. We had the talented group that could do most anything with very little money.

So, was mashing mud important? Yes, it was! David Beach was one of Bob Ginsburg's graduate students. David had observed what we were doing with lime mud for some time. When he graduated, he went to work for Amoco and soon became involved in a carbonate-exploration play in the Williston Basin in the Dakotas. The focus of the exploration—called a "play" in the oil business—involved a grainy-limestone reservoir surrounded by a muddy-lime unit that was relatively impermeable. The target, as seen on seismic profiles, was a porous topographic feature that stood higher than the surrounding muddy unit. It was a perfect setting for an oil reservoir. The problem was that company petroleum engineers and geophysicists could not explain why the grainy porous unit was higher and thicker than the surrounding muddy unit.

Conventional wisdom indicated faulting was the explanation, yet they could see no faults in the seismic profiles. They even tried to "wish in" some faults. David remembered our work and realized that the muddy unit could be lower and thinner than the grainy porous unit simply because it had compacted more. Limestone composed of sand-size grains is usually cemented early to form rock and is less deformable than mud. Faults were not needed.

David knew that when we had tried to mash sandy grainstones on Fisher Island, they did not compact. Why did the geophysics folks insist on faulting? Until then, they remained convinced that fine-grained lime sediment could not compact. Our compaction experiments, however, provided the simple and most logical explanation. There had simply been compaction of the relatively more compactible lime mud, and faults simply were not needed! Whether this helped them find oil or not I don't know. That would have been a company secret. I do know that David's new interpretation was a key element in the play. Just knowing that our experiments had practical implications in the real world was most rewarding.

Petroleum geologists were so interested in our results that the American

Association of Petroleum Geologists asked me to organize a two-day conference in Tulsa. The title of the conference was "Deep Burial Diagenesis." That title was posted on a sign outside our meeting room. I still remember being stopped in the lobby of the hotel and being questioned by a hotel resident. "Are you people undertakers? Is this some new way of burying bodies down deep?" The meeting nevertheless was a success.

So, does mashing mud have anything to do with whitings? Well, yes, but that is still to be discussed.

A New Journal

In 1986 the Society of Economic Paleontologists and Mineralogists decided to launch a new journal to be called *PALAIOS*. Its purpose was to highlight key fossils in the geologic record as well as the tracks and trails made by various organisms including characteristic burrows. Because of my work I was invited to be an associate editor, which I happily accepted, hoping to initiate a section in the journal devoted to new ideas and unanswered questions.

It seemed to me that science would move forward more rapidly if we attacked those aspects of geology that were least understood. To help get my idea across—call it a trial balloon—I published an essay in a section of the journal *ONLINE* (1.3, 1986). The purpose of my essay was to highlight a problem that had been gnawing at me for several years: that science got stuck in ruts because labs and funding tend to march in step with prevailing knowledge, which cuts down on new discoveries. I characterized the problem as "Paradigm Disease" wiping out "Gee Whiz Science." Sadly, the idea didn't catch hold. We still seem to be doing the same things over and over but with increasing accuracy. Looking back, I realize it was a brave essay because every geology department and research laboratory knew exactly which researchers I was criticizing. I didn't mean to be derogatory but just wanted to highlight a few things to make the point that our ingrained attitudes were holding us back from making new discoveries.

Cutting Steel with Water and the Bug Springs Adventure

Our little office/laboratory, isolated as it was from the mainstream of bureaucracy—a true Skunk Works—offered many advantages. We could cut bureaucratic corners and be of service to other projects within the USGS. One such project was for the Conservation Division, which in the future would be folded into what became the Minerals Management Service. (In 2010, after a devastating oil spill, MMS was reorganized and became the Bureau of Ocean Energy Management, Regulation and Enforcement, or BOEMRE.)

Because we were known for our diving and underwater work, some interesting and unusual opportunities presented themselves, frequently from a man named John Gregory who headed a program within the Conservation Division that contracted out research for new drilling and safety devices. It included everything from downhole sensors to underwater devices for cleaning platforms that enable safety inspections. He even had resources to fund our study of drill-mud plumes and also supported a graduate-student study to determine the toxicity of drilling mud on corals. We became good friends and supporters of John Gregory's program. He also was a bit of a maverick.

John Gregory had contracted with a private company in the D.C. area to develop a device for cleaning steel underwater. It was so powerful that it would not just clean; it would actually cut metal. The device was developed so offshore oil rigs could be cleaned for inspections that were needed to detect minute cracks in welds before they became a problem. The welds on offshore rigs are always heavily encrusted with barnacles and other marine growth, and cracks or fissures could not be seen because of the growth. Fouling organisms had to be cleaned, but how? A wire brush and a hammer, the traditional method along with motorized brushes, took much too long.

High-pressure water jets were the solution, but not the high-pressure cleaners available at your local hardware store. These consumer devices are used for cleaning sidewalks and for peeling paint and dozens of other objects. They spray garden-hose water at a pressure of around 1,200 pounds per square inch (psi). Some can be purchased or rented that produce up to 3,000 psi. The company that John Gregory had contracted to had made one that produced 10,000 psi, strong enough to cut steel and slice through aluminum as if it were butter. It was designed to be diver operated.

This is where we came in. The device needed testing in a real-life situation, and someone also needed to film the action underwater. We had these needs covered. We knew which contract boats were available nearby, and

I could do the filming with the 16-millimeter movie camera, and Harold Hudson could take still photos. There was another feature. With our location and knowledge, we could eliminate weeks or months of the contracting bureaucracy at headquarters. In no time, we had located a vessel large enough to carry the huge high-pressure pumps.

Within a half mile of our field station was a small lighthouse—a navigation marker to guide cargo ships into Government Cut and on into the Miami Harbor. We made a dive and determined that under water the lighthouse was constructed almost exactly like an offshore oil rig. It had the same kind of welded joints, and it was heavily encrusted—perfect for the needed test.

The test would require only one day to conduct. Harold and I accompanied the commercial divers to the channel marker. After anchoring nearby, we dove in to watch. The action of a 10,000 psi water jet is impressive. Ordinarily, it would be impossible to operate such a device. The diver would be jetted backward as soon as he pressed the trigger. To alleviate this action, the all-titanium device was outfitted with a jet that worked in the opposite direction. It countered the backward thrust of the cutting head.

I filmed the action and edited a short film that could be shown as proof the device actually worked. I never heard anything more about the fate of the device but assume the diving industry that services rigs purchased the jets from the manufacturer. Because we were in the Oil and Gas Branch, I concluded we were peripherally helping the nation's oil and gas needs. The Coast Guard that owns the channel marker never knew anything about what we were doing. Imagine the delays if we had gone through their bureaucracy! Ah, 'tis far easier. . . !

On June 3, 1979, a PEMEX rig drilling off Mexico on a prospect called Ixtoc experienced a severe blowout. Oil was soon washing up on south Texas beaches. Brown and Root, the offshore drilling and equipment supply company, improvised a device they called the sombrero, a funnel, or hat-shaped device, which was lowered over the runaway well. The floating oil would enter at the bottom and be funneled up the device to the surface where it was pumped into a floating tanker. The idea was good, but apparently the design wasn't quite right. It didn't capture all the oil.

The result of that attempt was a contract with the Massachusetts Institute of Technology (MIT) genius Jerry Milgram. Using his calculations, he experimented with and developed an improved recovery device. His device worked in a tank at the MIT laboratory, but a field test was needed. He found a place to test the device in a 150-foot-deep freshwater sinkhole in central Florida called Bug Spring, which was located in the middle of an orange grove. The Navy had already established a facility there for underwater-acoustics experiments, and it included a rather large floating dock and some onshore laboratory space for observation.

Milgram had arranged to test his quarter-size scale model in the spring. It consisted of a nearly ten-foot-diameter funnel turned upside down. At the small end was a six-inch-diameter pipe that led to the surface. The device had been lowered on cables down to about twenty to thirty feet below the surface, and a fluorescent dye was used to simulate petroleum. There were also bubbles of compressed air, simulating blowout gas that generated an upward flow of water. It reminded me somewhat of the airlift device used by treasure hunters to vacuum sand from a shipwreck. What was my role?

When all was ready, I descended to the entrance of the upside-down funnel and shot 16-millimeter movies. The films were to demonstrate the effectiveness and to show any unsuspected irregularities in its performance. I had to be careful to avoid affecting the flow of dye and bubbles and, of course, avoid being sucked into the funnel myself.

Several things impressed me. At first, Jerry seemed to be a stereotype of the absentminded egghead professor from a prestigious university. I really wondered if any of this would work. Was I surprised! That man could handle a welding torch one minute and operate a sophisticated spectrometer the next. At the same time, he was making complex calculations in his head. I was greatly impressed.

Part of the project involved a memorable trip to MIT where I filmed Jerry working with the laboratory model. While there, he showed me around MIT's Herreshoff sailboat museum. Milgram, it turned out, was also an accomplished sailor and designer of boat hulls. Years later, I read in a boating magazine that he played a role in designing an America's Cup sailing vessel.

You Came Here in That Boat?

There was, and remains, much concern about potential effects of the mud released during offshore drilling. Drilling mud is a mixture of various clays and other components pumped down the drill stem during drilling. The mixture lubricates, cools, and brings bits and pieces of rock called cuttings back to the surface. Geologists analyze the cuttings to determine the age and kind of rock being drilled. Offshore, some of the mud is discharged overboard, though most is continually recirculated. The discharged mud makes a plume that drifts down current and may persist for half a mile. What does it do to the environment? How much is there? These are just a few of the questions that were being asked.

With funding from John Gregory's Conservation Division program and using Captain Roy's older boat, the *Sea Angel*, we headed for the rigs off Louisiana and Texas. The crew consisted of Harold Hudson, Dan Robbin, me, and a student geologist named Carol Lee. We were all divers. Carol was on a scholarship, provided by the National Association of Geology Teachers. Over the years we had several of their high-achieving students, including Lisa Robbins, who worked on whitings and would one day be my supervisor.

We had some ideas about the drill-rig work that we would try on this trip. Before we left Florida waters, we stopped on a reef and collected healthy colonies of the staghorn coral *Acropora cervicornis* and kept them in two coolers on deck and changed the aerated seawater every few days. Our plan was to attach the coral along a thirty-foot-long underwater boom that would track in the current. One end of the boom would be attached to the underwater drill-mud outlet pipe. Corals closest to the pipe would experience more mud than those farther away. We would set up the experiment early, do other work, and return about two weeks later on our way home. We would measure how much the corals had grown, or not grown, and determine if there was a downstream gradient in growth rate. Earlier study showed that this species, under normal conditions, would have grown several millimeters during that time.

When we reached the area of rigs, we searched for one where drilling was in progress. It was easy to look up from our boat and see if the drill pipe was rotating or to watch for the mud plume. Unfortunately, every time we found a rig drilling, the outlet and mud plume were in the wrong place—in the shade under the rig. We needed one where the plume was moving away from the rig and in sunlight. We didn't find one that fit that description, so we proceeded with other work and left the corals in the coolers.

We had six Niskin bottles, which we attached at measured intervals along a hundred-meter-long plastic line. With this rig in tow we would approach the rig in a rubber boat, slip into the water, and attach one end of the line to the drill-mud outlet pipe. Being neutrally buoyant, the line of samplers played out in the current within the mud plume. The samplers were open at each end, so plume waters flowed through. When all was ready, one of us would swim down the line triggering the devices, sealing each end, trapping a ten-liter representative sample.

Back on *Sea Angel*, the ten liters of sample were pressurized and forced through a filter. The filters coated with particles were then stored for drying and weighing back at Fisher Island. They would reveal how much material was actually in each liter of water at different distances from the source. We also collected and filtered clear water upstream for comparison.

Curious roustabouts, tool pushers, and other rig workers generally looked down at us, wondering what we were doing. What did they think about divers doing strange things under the rig? This was long before 9/11, and terrorism was the last thing on their minds. Today, those Niskin bottles would be mistaken for bombs. We also had a trick up our sleeve. Rig workers who were not busy drilling or sleeping were generally strolling the decks or fishing. To avoid too much notice, we put Carol Lee on the front deck of the *Sea Angel* in a bikini. As we headed for the rig in the rubber boat, the *Sea Angel* circled to the other side. You guessed it! The men ran to the other side to get a look!

We sampled the plumes from seven different rigs using our clandestine sampling technique. On some occasions we were invited aboard. It was usually a Shell rig. I would explain how I had worked for Shell, and all were very helpful and often offered us food. What we heard in every case was, "You came all the way from Florida in that boat?" They were used to seeing large, fancy government agency research vessels. A government research vessel would undoubtedly have put them on good behavior. They certainly had no idea what we were up to.

The weather was good, so to round off the trip we headed for the Flower Gardens reefs. The Flower Gardens consist of corals growing on two separate salt domes about 110 miles offshore. The water around them was around four hundred feet deep, but the tops were as shallow as sixty feet. The domes are encrusted with huge coral heads and numerous fish so they attract divers—more than a thousand each year. I had been there in the early 1970s when I lived in Houston and was president of the Houston

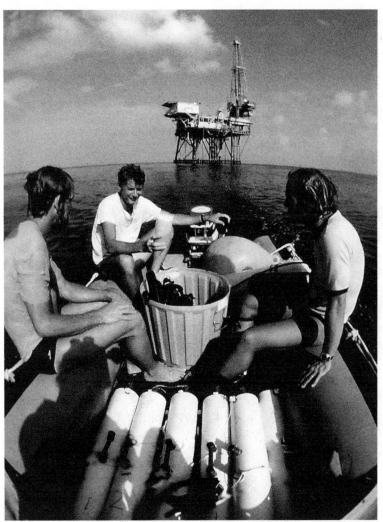

Investigating content of drill mud in the Gulf of Mexico using Niskin bottles.
Left to right: Harold Hudson, Gene Shinn, Dan Robbin. Photo by Steve Earley.

Underwater Club. That was when the reefs were first proposed as a NOAA National Marine Sanctuary. Sanctuary status was not bestowed until 1992.

We had our versatile hydraulic drill on board the *Sea Angel* so we could take some cores of large coral heads. We drilled the species with the best annual growth rings. The growth rings are much like tree rings and easy to count. In addition, bits of coral skeleton from known years can be analyzed for traces of drill mud. We were looking for a drill-mud additive called barite, a mineral used to make drill mud heavy. Barite is the same stuff you

drink before having a stomach X-ray. Barite, or barium, is often incorporated in coral skeletons. If present, you can count the rings and determine
the year in which it was introduced. We already knew the drilling history of
the area. We stayed at anchor for two days and took turns diving, collecting
a dozen two-foot-long core samples.

Finally, it was time for the long voyage home. It was September, and

Harold Hudson coring large-head coral to determine its age at a reef off Palawan
Island in the Philippines.

thankfully there were no hurricane threats. But what could we do with the staghorn corals still living in the cooler? They didn't look as healthy as they had earlier, but all were alive. Would they survive the trip back to Florida?

We had noticed a spot on the bottom near the center of the Flower Gardens reef, in sixty feet of water, where a series of lines came together. I was sure they were part of a research program being conducted by Dr. Tom Bright and other Texas A&M biologists. We decided to leave the corals at their experimental site. The biologists would be able to monitor their growth and health, although we expected they might die during the coming cold winter. After all, there were none living there—or were there?

What we did was, of course, anathema to biologists. We were introducing a foreign species where it did not belong. I figured they would die en route back to Florida anyway, so here was a neat experiment. Twice I had transplanted this species to Bermuda, and both times they died when winter cold came, so we knew why they were not native to Bermuda.

Upon return to Miami, I received a late-night call. It was via marine radio, and transmission was a bit fuzzy. It was Tom Bright. "Gene, did you leave some staghorn coral out here?" I said yes. "Do I have permission to remove them?" he asked. I said yes. And that was that! Or was it?

More than twenty years later, geologists Richard Aronson and Bill Precht, who were monitoring the reef for NOAA, found acres of dead staghorn and elkhorn coral at the Flower Gardens! They dated to around six thousand years old. So they were not exotics after all. Furthermore, the two scientists also found living elkhorn. Will someone soon find living staghorn? I felt vindicated.

The coral cores were X-rayed and the bands were measured. The cores were also analyzed for barium. None was detected. The surprise was that all samples showed that growth began slowing starting in 1957! That is still a mystery. Our findings from the drill-mud plume study and Hudson's data from the coral cores were published in a book less than a year later. It's amazing what can be accomplished with a bikini-clad girl on deck!

A Philippine Adventure

After the drill-mud study in the Gulf of Mexico, we still thought it important to evaluate the actual effects around wells drilled on coral bottom. One day a representative from Cities Service Oil Company approached us about such a study. He had heard our presentations at a drill-mud conference. We pointed out that no one had really looked to see the effects of drill mud on a live coral reef. We knew it was a vital and sensitive subject. I recalled my experience with the Australian Great Barrier Reef Commission. I knew it was an important issue.

The man from Cities Service said, "I have an opportunity for you. Our company has drilled on coral bottom off the Philippines. If we can get you over there, could you study the site?" "Well, yes," I said. We, of course, had no funds to travel so far, but it was suggested that the American Geological Institute (AGI) might find a study like this important. At the same time, the International Coral Reef Symposium would be taking place in Manila, and we wanted to go and make presentations about our work in Florida. After some negotiations, we made arrangements to have our airfare paid in order to combine the coral meeting with the proposed study. AGI, a nongovernmental organization, would pay our airfare and ship our equipment and in addition pay for an underwater photographer to document our work. I suggested Steve Earley.

It was a complicated undertaking, but we flew to Manila, first class, and with all our drilling equipment in the luggage compartment. At Manila a company representative met the plane and whisked us away to a tanker off Palawan Island. We would be living on a moored oil tanker not far from the actual work site. There was also a nearby producing oil rig that was filling our moored tanker with crude oil through a single-point mooring system. A few miles away from the producing rig was the recently drilled site in eighty feet of crystal-clear water. A buoy tethered to a heavy chain marked the site. The actual drill hole consisted of a thirty-foot length of twenty-four-inch-diameter casing protruding upward from the bottom, waiting only for a platform to be installed over it.

The site was not exactly a coral reef, but many small-head corals were scattered over the seafloor. A brown coating of drill mud had accumulated on the bottom surrounding the vertical casing. The coating of mud was obvious, so it was clear that all the corals in the area had experienced an abundance of drill mud. The damage that was most obvious was that caused by the chain holding the buoy. The chain had dragged round and round forming a circle of scraped bottom. There were no corals remaining in the scraped circle. Our work would be outside the circle.

Each morning, a crane on the tanker lowered a Boston Whaler outboard motorboat with all our equipment. The tanker was large, so it was about thirty feet down to the water. When our boat was in place, we were lowered in a cargo net. My job was to get the motor running and head to the study site. This study would be Harold Hudson's baby. The rest of us—Dan Robbin, Steve Earley, and I—would be his assistants. We were accompanied by a visiting Korean geologist named Dong Choi, who was working with Bob Ginsburg at Fisher Island. Choi would be studying one of his specialties, the microboring organisms that were living in the various rocks we brought aboard. He was also presenting a paper at the upcoming Coral Reef Symposium in downtown Manila.

Once again, Harold used the hydraulic coring device to take short cores in living coral heads to see if the drilling mud had deleterious effects on growth. The corals were not nearly as large as those at the Flower Gardens. We knew there would be barium in these corals and that the well had been drilled two years earlier, so it would be possible to compare recent growth of the skeleton with that from before the drilling.

We had a terrifying experience halfway through the study. One afternoon we returned to the tanker late in the evening. The crane operator lifted us aboard in the personnel basket as usual. The small boat was tied alongside, but the stern was facing toward the wind and waves and water began to lap over the transom. As we watched in horror, our boat, with our equipment, filled and turned over! We were helpless. Because of the language barrier, we couldn't communicate with the crane operator. The water was at least two hundred feet deep, well beyond our diving depth, and it was already dark. We turned away in disgust and retreated to our cabin. There was nothing we could do. How were we going to go home empty-handed?

About an hour later, someone came to our cabin and said, "We got your boat and stuff on board." How could that be? We had seen it sink. As it turned out, the crane operator simply pulled the boat up by a single line. It was upside down, but fortunately we had tied the hydraulic power source to the boat's steering column. It was just a short length of old sisal line, but the hydraulic pump remained tethered to the boat as the crane pulled it aboard. Attached to the hydraulic pump were the two hundred-foot-long hoses that were also attached to the drill. The quick-release fittings had held! The workers had dragged all the equipment and the boat back on board. The cheap quarter-inch line attached to the steering wheel had saved it all!

By noon the next day we had drained and flushed all the saltwater from

Gene and Dan Robbin taking a core of live coral near an oil well in eighty feet of water off Palawan Island in the Philippines, 1981. Photo by Steve Earley.

the gasoline motor. We dried the electronics and replaced the hydraulic oil and were soon back in business. It seemed like a miracle!

We finished up the work and soon were ashore and headed to the International Coral Reef Symposium in downtown Manila. Thus far the entire trip had been a success, and we all presented papers. Harold's results from the well-site study were later published in the *Bulletin of Marine Science*.

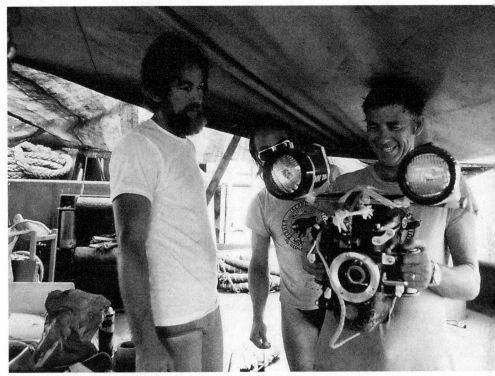

Gene with underwater movie camera aboard a tanker off Palawan Island, Philippines. Harold Hudson (*left*) and Dan Robbin look on.

Steve Earley had taken great underwater photographs, and we even survived the poisonous lionfish that were everywhere. What was our major conclusion? The chain attached to the marker buoy had caused nearly all the coral damage! Drill mud had not affected the growth of corals that were clear of the grinding anchor chain. That was quite a surprise! We didn't realize that this would not be our last look at the effects of offshore drilling. A more contentious study lay ahead.

There was one other great side trip while in Manila. We visited Corregidor at the mouth of Manila Bay. Corregidor had been General McArthur's headquarters before the Japanese invasion of the Philippines. He had escaped in the nick of time, but those who remained would eventually endure the well-known 1942 Bataan "Death March," during which thousands of prisoners would die. We could see the Bataan Peninsula not far away. It was a somber educational experience.

Winding Down on Fisher Island

In the late 1980s, funding for USGS research was dwindling. Research funding related to oil and gas geology declined with the drop in oil prices, and production shifted overseas. Once again we were becoming dependent on foreign energy. Petroleum geologists were being laid off everywhere in the United States, and geologists with PhDs found themselves driving taxis! The suicide rate among petroleum geologists, especially in Denver, increased sharply. "Only a pigeon could make a deposit on a Mercedes" was a common joke in Denver. Some universities dropped or reduced the size of their geology departments. In some cases, they merged with agriculture departments or became a new discipline called "Earth Science." Even jobs in environmental geology, usually involving the detection of runaway underground gasoline and other spilled products around gas stations, were waning.

To save Fisher Island Station, we transferred out of the Oil and Gas Branch and into the Marine Geology Branch, where there were more resources and greater public support. Thanks to Jacques Cousteau, anything related to the oceans had become more popular. We jokingly called it the "Cousteau effect."

We were fortunate to make the transfer. The key person who masterminded our transfer to Marine Geology was Bob Halley, one of the Fisher Island originals. Bob had moved through the system and become chief of the Marine Geology Branch in Woods Hole. It was a good move for both Bob and us because it meant that our research no longer mandated studies related to oil and gas. There were many environmental issues we could help solve, and besides we knew the terrain very well. Harold and I knew every nook and cranny of the Florida Keys reefs, and we also had the core drill!

The new assignment provided many and varied adventures. We began by mapping and coring in the NOAA National Marine Sanctuary in the Florida Keys, but this time it was for different reasons. We still wanted to understand coral reef distribution, and NOAA was interested enough in coral distribution to pay for much of our research. Corals were beginning to die, and we and the managers at NOAA wanted to know why. We also conducted a related study to evaluate environmental effects of offshore oil wells. This time we would examine sites where exploratory wells had been drilled off South Florida. The Minerals Management Service (part of the

Interior Department) came to our aid and funded the study. We were also free to study groundwater contamination, which was becoming a large issue. In addition we would tackle the potential effects of "Mount Trashmore," the mountainous pile of garbage and soil on the southern shores of Biscayne Bay. We even moved farther afield to the Marshall Islands a few thousand miles out in the Pacific Ocean. There were many new forks down the road waiting to direct our activities.

Big Bang Adventure at Enewetak Atoll

Together with Bob Halley, Harold Hudson, and Jack Kindinger, who came to us from the USGS Texas Corpus Christi office, we began what would be a two-month adventure in the Pacific. We would be using our diving skills and well-traveled underwater drill to examine the effects of hydrogen bomb craters. This time, the Defense Nuclear Agency would pay for the work. They had already initiated a seafloor and subbottom survey conducted by geologists from the USGS office in Woods Hole.

To get started, we first had to transport a mountain of equipment, including our faithful twenty-five-foot research boat *Halimeda*, to San Diego, where we loaded all of our equipment aboard what would be our new home. The vessel was a converted offshore supply boat named the *Egabrag*. The boat's owner had prospered hauling garbage in the Marshall Islands—and he had a fine sense of humor. The boat was uniquely named "garbage," spelled backward.

Later we flew from Miami to Kwajalein, the largest atoll in the Pacific. Kwajalein had experienced terrible fighting in World War II. It was, and still is, where the Air Force tests missiles and other secret devices. The missiles are shot from Vandenberg Air Force Base in California and land in the lagoon at Kwajalein, where divers recover the remains. Meanwhile, Russian spy boats disguised to look like fishing trawlers would observe from offshore. One such boat stayed visible just offshore the entire time we were on Kwajalein.

At Kwajalein we boarded the *Egabrag* to begin the two-day cruise over to Enewetak Atoll. *Egabrag* would be our home for the next two months. During the day we would be scuba diving and drilling from the *Halimeda*. At times we did our work using a research submarine. *Egabrag* had an amenity that helped cure my homesickness. It was the captain's hobby—he was a radio ham. Around 2 A.M. I would wake up and stagger to the bridge and call W4BIC, the call letters of my father's ham rig back in Florida. When I was at Enewetak, my family had already begun to call EK Papa-San, the name he adopted from servants during his duty in the Far East. It was wonderful to talk with Papa-San, and sometimes transmission was so clear it was as though he was in the room. I could hear his dog bark.

Besides our drill, we had brought with us a new research tool—a little yellow two-person submarine named *Delta*. The sub, piloted by my long-time friend Richard Slater, allowed us to map and collect on the bottoms of two-hundred-foot-deep nuclear bomb craters. Mapping from *Delta* became Bob Halley's main project. The best thing about the submarine was that abundant highly aggressive reef sharks couldn't get in. The little yellow submarine *Delta* was fantastic and never required maintenance!

Ironically, the project was called the PEACE Project, which stood for Pacific Enewetak Atoll Crater Experiment. A project called PEACE to study the effects of hydrogen bombs! Does that sound political? Actually, the

R/V *Egabrag*, our Enewetak home for two months, launches research submersible *Delta* in an atomic bomb crater while Boston Whaler stands by for communications.

Harold Hudson and Bob Halley drilling at edge of Oak crater as submarine *Delta* cruises by. The crater is two hundred feet deep.

entire study was politically motivated—as is much government-sponsored science—but that didn't concern me at the time. The political background is this. Congress was at odds over a military proposal to install MX missiles in the United States, the basic question being whether to pack them close together in cement-lined silos, the "dense pack" system, or to scatter them farther apart, that is, "loose pack." A third possibility was to deploy them on railroad cars that would keep moving around the country to confuse the enemy. This was during the Cold War, before the Berlin Wall came down, so we all knew who the enemy was. Surprisingly, after all the many dozens of tests of atomic devices in the Marshall Islands, the precise crater size a weapon of known yield would make was poorly known.

The reason that information was not known is because the size of a nuclear crater enlarges within a few days after "the event," which is military talk for an explosion. For several days after the events, radiation is too high for humans to make accurate measurements, and within those several days the craters get larger due to subsidence. They enlarge because the underlying rock is crushed hundreds of feet below ground zero and the ground keeps caving in. Crater size at the moment of the event was the information

Hydrogen bomb craters on the reef at Enewetak Atoll in the Marshall Islands, 1981. Mike was the world's first detonated hydrogen device, and its crater is overlapped by Koa crater in the foreground. Both Koa and Oak were plutonium devices. Oak was the last test of a plutonium device at Enewetak, with a yield of nine megatons. The Oak crater (in the distance) is two hundred feet deep. Photo by author.

Gene videos coral growing on railroad rail that was part of the subsided road to the Koa atomic crater at Enewetak.

needed to make a decision about how to harden and space the missile silos. Our job was to determine what the size had been at the instant of the explosion. That involved some keen observation and geological knowledge.

We were working at Enewetak Atoll because it was the only place where large U.S. hydrogen devices of different yields had been tested above ground. Previous surface tests were of small devices, and subsequent tests had been conducted far below the surface in Nevada. The Russians, however, had shot many large bombs on land and most likely already had the kind of information our military needed. Unlocking a crater's secrets would require a lot of geological research. Fortunately for us, we had a secret weapon. We had Gene Shoemaker! Gene was the scientist who predicted the impact of comets on Jupiter and had made the cover of *Time* magazine. For many years he had traveled the planet measuring various craters made by stony visitors from outer space.

Gene had worked on the well-known Meteor Crater in Arizona as well as numerous others. He had also conducted many studies of the smaller atomic craters at the Nevada test site. So with all of us wearing moon boots, Gene led us into some nuclear bomb craters at the Nevada testing grounds. With Gene, we also clambered all over bone-dry Meteor Crater so he could teach us what to look for underwater at Enewetak. The key to determining the crater size, whether a comet or a nuclear bomb made it, was the "overturned flap." Basically, the layers of rock at the edge of the crater are raised and sometimes tilted back away from the crater to form a "flap." The entire episode from Nevada, to Arizona, to Enewetak, turned out to be a most interesting and unforgettable assignment. We made observations that impressed the military.

Interestingly, by the time our work was completed, Congress had moved on to other more important interests. The whole MX debate, like old soldiers, just sort of faded away. The Cold War was about over, and nuclear submarines with multiple missiles and multiple warheads became the preferred deterrent. Eventually, the wall came down.

The military folks also encouraged us to do any basic research we wanted while we were there. For me this was an excellent opportunity to broaden our understanding of marine cementation around reef margins. I would also have an opportunity to examine spurs and grooves in a setting different from Florida. Spurs and grooves had been my first objects of study in Florida back around 1960 and led to my first scientific publication. Dynamite wasn't needed this time!

We knew from other studies and the work I had done on Cretaceous reefs back at Shell University that the process of cementation can fill the pores in marine sediment. The process of cementation is more intense near reef margins, where pounding waves continually force seawater through the rocks. Clearly, seawater was the vital part of the cementation process, and Enewetak presented the perfect opportunity to core-drill near the outer margin of a classic atoll. We were able to set up the drill next to pounding Pacific waves, where wave action was so intense and dangerous that we often had to back off and wait for low tide. We drilled a line of boreholes leading away from the outer reef-flat margin and toward the center of the atoll. The drilling revealed a dramatic decrease in cementation and rock hardness as we moved toward the lagoon. The rock was so hard near the

Core drilling near heavy surf near the reef edge at Enewetak.

margin that we broke the teeth off the drill bit. Farther away there was only soft sediment.

Diving in the grooves between spurs was exciting. The major problem was how to enter the water when there were crashing waves. We could do it only on calm days, and even then we had to time our entry between incoming waves to avoid being bashed against the rock. What were very different from the grooves in Florida reefs were large, rolling boulders. Wave action is more intense in the Pacific, and coral boulders were nearly always rolling back and forth. There were also potholes with smaller boulders that whirled around in the wave-induced current and enlarged the holes. It was like what one often sees in streambeds. So what did I learn? Much of what I had reported about Florida reefs did not hold up at Enewetak. In this case, erosion seemed to be a major creator of the spur-and-groove system. In Florida, it was coral growth.

Defense Lingo

All of us learned some interesting military jargon at Enewetak. "Devices" was code for bombs. And devices did not explode, they produced "events," and of course we began using terms like "ground zero." One of the craters we worked in was made by a device calculated to be equal to about ten megatons of TNT, and the "event" had created a crater two hundred feet deep. And then there was "Mike." Mike made a huge crater. Mike, the first hydrogen-fusion device, was the size of a small warehouse and used heavy water (deuterium) rather than plutonium. It was Edward Teller's dream bomb. Teller had long lobbied for development of a "fusion device," that is, hydrogen bomb. The first atomic bombs had been "fission" bombs, which used uranium. They were less efficient than fusion devices and produced more radioactive fallout.

We found it interesting that the radiation-monitoring officers, called "rad men," carried Coleman-lantern mantles in their wallets. Why? When they wanted to see if their Geiger counters were working properly, they would simply place the lantern mantle over the detector. We learned that lantern mantles contain radioactive thorium. The amount is small, but ironically it was just enough that if it had come from a nuclear power plant it would be subject to the same regulatory process as all other nuclear waste. It made me wonder about all those antinuke protesters camped at night around the gates of nuclear facilities with their gasoline lanterns blazing. If they only knew!

There was one negative outcome of the Enewetak assignment. For several years we had attempted to have Dan Robbin's title changed to geologist. Such a change within the government system was not really in our hands. We had to negotiate with headquarters in Reston, Virginia, and we had tried repeatedly and failed. There was little long-term USGS future for Dan by remaining a physical science technician, and the prospect of working in nuclear-bomb craters was a little frightening for him. He saw his fork in the road and took it. Here was his chance to go back to school, so Dan enrolled in a psychology program in Miami.

Today Dan is a licensed social worker and programs coordinator with the Veterans Administration Hospital in Miami. He regularly treated Vietnam vets, and as times changed his patients became Iraqi and Afghan war vets. He has told me that because of roadside bombs, these veterans have larger and more difficult emotional problems than did the Vietnam veterans. Dan makes regular inspection tours throughout Florida and visits us in St. Petersburg often. He had taken the right fork.

A 3.5-Billion-Year-Old Mystery

He was very excited when he called! It was my longtime friend Bob Dill on the phone—it was 1984.

Bob and I had met in 1964 during a Geological Society of America field trip in the Florida Keys. Bob Ginsburg had arranged the field trip—I was helping. Dill was a well-known diving pioneer who had been the first diver to reach the wreck of the Italian cruise ship *Andrea Doria*. On a foggy night in 1956, the ship collided with a Swedish freighter and sank in over two hundred feet of icy-cold water off New Jersey. Dill was featured in an underwater photograph on the September 17, 1956, cover of *Life* magazine.

Dill had done other exciting deep diving, both with scuba and in submarines. He had conducted his PhD research in the depths of Scripps Canyon off southern California while working with Francis Shepard and was also one of the first U.S. diving scientists to work with Jacques Cousteau. Over the years we had been involved together in many projects. He had wrangled my ten-day voyage on the *Calypso*, where we helped in the making of a television special on blue holes. We made many dives in blue holes and we trusted each other, especially under water.

There was tangible excitement in Bob's voice! "Gene! I have found submarine stromatolites in the Bahamas," he blurted out. "You need to get over here quick and take a look." I said, "But Jeff Dravis found some in the Bahamas last year. He published his find in the journal *Science*." "But those were small and in only two feet of water," he replied. "These are different. They are six feet high and the water is twenty feet deep!" Indeed, they were!

Stromatolites are sedimentary structures constructed by a unique symphony of sand and cyanobacteria. Stromatolites and cyanobacteria were the first forms of life on Earth more than 3.5 billion years ago. Being the first microscopic plants, the cyanobacteria that constructed stromatolites consumed carbon dioxide that dominated the atmosphere and expelled oxygen. The oxygen they expelled is believed to have allowed the first animal life to evolve and eventually dominate the Earth. What Dill found were modern forms of what had once been the most abundant life on Earth.

Adam Ravetch measures heights of living stromatolites discovered at Lee Stocking Island in the Bahamas by Robert F. Dill. Photo by R. F. Dill.

There is only one other known place where they live today—the salty ponds called Shark Bay on the west coast of Australia.

The reason stromatolites thrive in Shark Bay is because the water is so salty that fish, snails, and worms that eat algae cannot live there. The algae that produce them have no predators in Shark Bay. Anywhere else in the ocean, oxygen-breathing predators eat the algae as fast as they grow. What

Bob had found were truly giant forms, some more than six feet tall, living not in very salty water but in normal ocean water! Here they were protected from predation by viciously strong tidal currents. These giants live in a tidal pass next to Lee Stocking Island in the Bahamas. The currents are so strong they can be observed underwater only at slack tide, which lasts for only about ten minutes. Ten minutes isn't enough time for predators to find and nibble all the algae responsible for their growth. Fortunately, there was a marine laboratory within a stone's throw of the stromatolites.

The Lee Stocking lab was the brainchild of John Perry, who owns the entire island. Bob had been making many trips to the Caribbean Marine Research Center (CMRC), where he worked with Bob Wicklund and his wife, Jerrie. As a team, Bob and Jerrie managed the laboratory, which was partially funded by the NOAA Undersea Research Program. This is the same Perry who is well known for making underwater habitats and submarines. At CMRC, Perry was investigating alternate forms of energy. There were solar panels and windmills, and he was even making alcohol from seawater! At least that's what we were told. The secret devices were in a sealed container like those carried on container ships. The laboratory boat actually ran on the alcohol they made there at the island! It was indeed expensive alcohol and probably not economical for widespread use, but it nevertheless ran the research vessel. It was admittedly a cranky engine to get started. Besides managing the laboratory and associated facilities, Wicklund was raising native conchs and experimenting with adapting freshwater fish to live in saltwater. They were projects aimed at feeding third-world countries.

I soon found myself flying to Lee Stocking to see what Dill had found. About twice a week John Perry's private plane made the trip from Palm Beach to Lee Stocking Island, where it lands easily on the island's three-thousand-foot runway. When there was room, the flights were free! Thus began a long period of frequent visits to Lee Stocking. Even better, food was provided and there were no costs, so no government travel vouchers and paperwork were involved. That way we avoided the task of obtaining State Department permission to work in a foreign country.

What I saw when Dill took me diving was spectacular. It is amazing that these features had been there for so long without being noticed. There were rows and rows of the structures. We began calling them castles of sand. Why? They are composed of ooid sand bound together by microscopic plants and cemented by aragonite.[5] It was yet another example of submarine cementation like the process of rock formation I had found in the Persian Gulf.

Before long, we had Chris Kendall and Randy Steinen visiting with their students. Dill's only disappointment was that he could not obtain funding from the NSF to more thoroughly study his find. We did all the work without official funding.

Dill and Wicklund submitted several research proposals, but none were funded. It was yet another example of the politics within science funding. We were not known as experts in the field of stromatolites, although we learned fast. The highlight was that we published the discovery in the journal *Nature* and made the cover of a 1986 issue with a spectacular underwater photograph. And of course there followed a series of news articles, radio talk shows, and presentations at national meetings.

The discovery helped elevate the reputation of the Caribbean Marine Research Lab and Fisher Island Station, which were badly needed. The USGS was experiencing reduced funding and possible closure of Fisher Island. Dill and I also prepared an educational 16-millimeter film called *Castles of Sand* that became popular at geological conventions. Nevertheless, the times were changing! To keep going, we turned to some unusual studies to seek funding from other agencies.

A Contentious Study

One of the contentious issues in Florida, which remains so, was the prospect of offshore oil drilling. The results of our drill-mud study in the Gulf of Mexico and the study in the Philippines had little effect on decisions to drill off Florida. The controversy raged on.

Before the closing of Fisher Island, I attended a public hearing on offshore drilling in Key West. As part of attending the hearing, my wife and I had decided to combine the hearing with a dive trip for Jim Ray, the fellow I helped recruit to take my place before leaving Shell. He was scheduled to testify in favor of drilling at the hearing.

On the day of the hearing, Pat and I met Jim in the lower Florida Keys with my boat and took him offshore to Looe Key Reef. We would do a little scuba diving and photography and assess the health of the reef. On the way out, Jim looked in the distance and said, "You don't have any drilling down here, do you? What is that?" He was looking at the iron lighthouse on American Shoal reef. That comment stuck with me. I realized that if the lighthouse were off Louisiana or Texas, people would think it was just another drilling rig. They were used to structures like that, no problem. On the other hand, people in Florida—and in California—were conditioned to view rigs as ugly and evil. Ironically, offshore lighthouses in Florida are symbolic objects of beauty with great historical value. What a difference in attitude a few hundred miles can make!

That night was an event to remember. A few hundred people assembled in the school auditorium on Stock Island near Key West. About sixty-five people testified, and there was electricity in the air! It reminded me of a religious tent revival. Of the sixty-five who testified, only about five favored drilling. Those against put on an interesting—often entertaining—show. One was proud to announce that her grandfather had engineered and built those historic lighthouses. Lighthouses were cultural icons, but she was dead set against drilling. Rigs are ugly! I remembered how earlier that day a lighthouse had been mistaken for an offshore oil platform.

A most memorable testimony came from a grizzled and bearded character carrying a bucket. I feared it was something nasty to dump on the government representatives seated at the long table. He launched into a story that went something like this. "I came to Key West twenty years ago in search of truth! During the last twenty years, I have learned that there is only one great truth. Eventually everything leaks." The applause almost brought the house down! Meanwhile, I concentrated on the government representatives at the front table. They were enduring all that pent-up animosity, and I couldn't figure how they could stand the abuse. They were called everything under the sun, their manhood was openly questioned—an ugly scene it was. This was Key West at its worst, or best, depending on your point of view.

Fortunately, I knew two of the men at the table from my Shell days in Houston. They were representing the Minerals Management Service, the Interior Department agency in charge of conducting offshore leasing. During a break, I walked up to the stage and talked with them, and I mentioned that I had personally flown over an oil well being drilled near Key West in 1959. I knew about where the well site was located. "Why not study and document the effects of that drilling? We all knew the world didn't come to an end because of it!" I pointed out that most of the people in the room either didn't live in Key West when the well was drilled (there were, in fact, fourteen wells drilled in the Keys) or they hadn't yet been born. I could see they liked the idea.

The hearing resumed, and it was Jim's turn to testify. We agreed we would not sit together afterward. Later we would meet in the parking lot. It was not safe to be seen together. We had a wonderful meal by the water later that night. That was my first experience at a U.S. public hearing. My experience at the Australian Barrier Reef Commission hearing was nothing like this!

A week later, I followed up the discussion with the MMS representatives and sent them a letter proposing a study of drilling effects on the coral reef environment. In a few weeks I was contacted and promised modest funding to search for the well sites. MMS provided the latitude and longitude for the wells, but the wells had been drilled long before the advent of the GPS system. We did have Loran C. Finding the sites depended on how accurately the locations had been plotted in the first place. The wells had been drilled before Loran C was established, so locations had been determined by triangulation between towers placed at fixed points on land. We had no idea how accurate the coordinates would be.

The search itself was a wonderful adventure. One site was very near where famous treasure salvor Mel Fisher would find the Spanish treasure ship the *Nuestra Señora de Atocha*. It was a galleon from which he would recover around forty tons of silver bars. There were also emeralds and rubies and other precious treasures, not to mention gold bars and other trinkets.

Captain Roy was eager for the wellhead search. He too had been a treasure hunter, and searching for the well sites was just like searching for Spanish galleons. Roy put us in contact with a treasure-hunter friend who was also an agent for a British company that made proton magnetometers. The instrument was inexpensive and was the same kind that Mel Fisher was using for finding treasure. Geophysicists at our Woods Hole office said it would never work. The device didn't cost ten thousand dollars like those used for scientific research. We paid about nine hundred dollars for the instrument.

We did have something going for us. We knew that when those wells were drilled the drillers would have thrown a lot of stuff overboard. A 1960 magazine article about the drilling of one well off Key West said, "CORAL CAUSES DRILLING PROBLEMS." What they meant was the substrate was soft. Today, that statement would mean something entirely different. If a lot of stuff had been thrown overboard, the sites might be detected with our inexpensive magnetometer.

We also knew that to drill through the so-called boulder zone and not lose all the drilling fluid, the drillers had to install at least three thousand feet of iron casing. They then drilled down through the casing. When the drilling was finished, the casing was cut off at the seafloor and left in the ground. As it turned out, a vertical iron body that large makes a very pronounced magnetic anomaly.

We converted the latitude and longitude data provided by MMS to Loran C numbers and began the search by placing a large buoy at the supposed drill-site coordinates. We began searching in a back-and-forth grid. It was like mowing the lawn, as treasure hunters call it. When we got magnetometer hits, we threw smaller buoys in the water. It was not long before we had a major "hit" caused by the vertical pipe in the bottom.

With buoys marking the spot, we would anchor the boat and dive in. There were no gold bars, but the drill sites were obvious. Casing was not visible, but indeed there were literally tons of iron and old cables scattered about. We documented it all with video. We had found the first site, so we used the same methods to find another in seventy-two feet of water and yet another in fifty feet of water. Two more were in about fifteen feet of murky water near the Marquesas Keys west of Key West. One of those was the one I had flown over many years earlier when it was being drilled.

The MMS office was very pleased when they saw the video. Not long after, they funded the rest of the study. It was at this point that I once again ran up against the politics of doing science. This study would likely have meaningful societal and political implications, which meant it was like waving a red flag.

I had known Walter Jaap, the state of Florida's official coral reef expert, for many years. We had talked about such a study, and he was eager to join. The politics of having a state official involved with the work seemed logical, especially because a federal agency would be footing the bill. It looked like a good deal for all concerned! I was wrong!

Walt was an employee of the Florida Department of Environmental Protection (DEP) in St. Petersburg. His supervisor was Karen Steindinger, who was also the country's best-known expert on red tide. (The name of the offending organism, *Gymnodinium brevis*, was later changed to *Karena brevis* to honor her groundbreaking research.) Karen was amenable to conducting further drill-site studies, especially as there was outside funding. However, she anticipated a problem. The governor of Florida, Bob Martinez, was publicly against all offshore drilling, and the DEP was very much under his control. What if our study showed that the drilling impacts were minimal and not long lasting? How might that affect the DEP? Karen took the logical action of most any agency leader. Walt could not be involved in a potentially controversial study that might offend the governor. Now what to do? This study needed a real biologist with credentials, not a bootstrap geologist like me.

It was at about this time that Phil Dustan, the same biologist who had worked on coral reefs for Harbor Branch, had become a new young professor at the College of Charleston in South Carolina. We had been good friends over the years. He was also the same person who would later influence me to engage in the study of African dust!

It so happened that Phil had no funding or exciting project on the horizon, and he loved diving. Would he do the biology? The answer was yes! In fact, he was quite excited about the prospect. We would be diving every day, and later in the project we would be using a research submersible and work on some newer and deeper sites in the Gulf of Mexico.

We completed the study and, much to everyone's surprise, found little lasting harm. All the sites had recovered, and at some sites the debris left behind had produced artificial reefs, or as Phil called them, "oases in the sea." Phil was lead author of the study results that we published with Barbara Lidz in the *Bulletin of Marine Science*. I prepared the in-house report for MMS.

I learned a lot from that study, and we were all very proud of our work. No one had ever performed such a study in Florida, and we really wanted people to know what we had accomplished and to realize the significance of the results. Government reports are generally called gray literature and are not well distributed. We wanted a wider audience, so I convinced MMS to pay for twelve hundred reprints of the peer-reviewed journal article. We were certain it would be an important and popular report. After all, impacts of drilling for oil and gas were the hot topic of the day and remain a hot—sometimes an overheated—topic!

Surprise of surprises! We received only about twenty requests for reprints when we expected hundreds. It was another lesson about science and scientists. The effort and quality of our work didn't seem to matter. What mattered were the expectations of the readers. That experience taught me that peer-reviewed journal articles are not always what they are thought to be. The problem was clear. We didn't find and document the widespread long-lasting death and destruction that was expected! We didn't find what the public, the media, and the academic community wanted. Isn't it death and destruction that sell newspapers? I think Mark Twain once said something like that.

I can make these statements because experience had taught me that whenever I published some small paper on coral reefs that had no societal implications, about a hundred requests for reprints would arrive by mail, sometimes even before I saw the published article myself. That's the way it was before the Internet. The Internet has changed everything. Reprints have little value now because most everything can be found on the Internet. Why clutter up your office with a reprint collection?

In addition to reports, we also produced a fifteen-minute underwater video showing our findings. It was distributed as an official USGS Open-File Report. This was something the public and politicians could appreciate. They could see for themselves what was underwater. Wrong again!

MMS, however, was very pleased with our study. In fact, that study led to yet another. This time we would examine recently drilled sites off the west coast of Florida where some thirty-eight offshore wells had been drilled. One was drilled just twelve months before we examined the site. We would examine and collect sediment at six different sites, including one in over four hundred feet of water. We now had GPS, and this time we would use the two-person submersible *Delta*, the same sub and crew we had worked with in the bomb craters of Enewetak. The study would involve launching the submarine from the Florida Institute of Oceanography research vessel *Suncoaster*, a state-owned vessel.

Again a report was published by MMS, and we prepared a much more professional video that was also distributed by the USGS. This time I decided that publishing in a peer-reviewed journal would be a waste of time and money. I was learning. We did document more impact but not a great deal of destruction. I suppose it depends on one's interpretation of destruction.

Basically, we found that if there was an abundance of junk at the site, there were more fish. In fact, the fish helped us find the sites. When the submarine reached the bottom, we would look for remnants of electric welding rods. They were good indicators that we were in the right place. Next we would look for fish. When we saw fish, we followed. They invariably led us to the borehole. Boreholes were generally full of fish, including large groupers. We found many small man-made objects—shovels, deck chairs, buckets, mops, cables, rolls of duct tape, and piles of drill cuttings. Some sites had practically no debris, but there were always fish clustered around the drill hole.

What was the outcome of all these efforts to document environmental effects of offshore drilling? As near as we could can tell, none! Our results had absolutely no influence on decisions to drill or not drill. In 1990, Congress provided the state of Florida with a moratorium. No drilling within two hundred miles of its shore. Eighteen years later, in October 2008, the moratorium was allowed to expire, most likely because oil prices had exceeded $150/barrel, gasoline had reached four dollars a gallon, and diesel was over five dollars a gallon. At the same time, at least 60 percent of the U.S. supply was still being imported from foreign sources. All this reinforced what I had already determined—and why I had left Shell so many years earlier—drilling decisions are emotionally and politically based and influenced by people and environmental organizations with little interest in scientific evidence or the welfare of our domestic petroleum industry. They just want to drive their cars. Only the high price of gasoline matters.

After the second study of deepwater well sites, completed in 1993, I decided not to attempt further work of this kind. I would soon begin examining groundwater and sewage movement under the Florida Keys. Unlike the oil-drilling results, the groundwater results were well received by the public, scientists, and politicians alike. However, our study of a landfill locally known as "Mount Trashmore" made few friends.

Mount Trashmore

With the "nuclear devices" experience behind us, we were soon drawn into another controversial area of science. It had to do with garbage. Some colleagues at the Florida Sea Grant Program in Miami had become concerned about possible environmental effects of a huge landfill on the shore of south Biscayne Bay. The landfill in question was, and remains, the highest point of land in South Florida and can be seen from almost any part of Biscayne Bay. It's the South Dade Landfill at Black Point, known locally as Mount Trashmore.

Fishermen suspected that it leaked some bad stuff into the bay, which is a part of Biscayne National Park. We were not funded to do a study, so one day we decided to go fishing and simply loaded up the drill and took it out on the salt marsh and mangrove swamp adjacent to the two-hundred-foot-high mountain and core-drilled. We had not installed water-monitoring wells before, so this was our first attempt. No one really knew what we were doing, and we were careful not to ask for permission. The government license plate on our vehicle kept most curious onlookers away. We looked official. Remember our motto: "'Tis easier to seek forgiveness than to seek permission."

We had no idea what we might catch. We really were fishing. Also, we were not really violating any laws that we knew of, but we knew full well that if we asked any agency, all kinds of paperwork barriers would materialize. If we did get permission we would have been watched like hawks, so we just quietly drilled and installed four shallow water wells, including one that was thirty feet deep. It didn't take rocket science, as they say, to see that we had stumbled onto something important!

The groundwater in a highly permeable zone smelled and looked awful. It was fluorescent green and smelled a little like turpentine mixed with ammonia. After collecting water samples from each well, we managed to obtain an emergency grant of one thousand dollars from the Florida Sea Grant organization. The money was provided because it was one of their agents who suggested we take a look in the first place. We used all the money to pay for pesticide analyses, which are complex and expensive. Just what was in the foul mixture?

Dr. Gene Corchran, an emeritus professor at the University of Miami Rosenstiel School of Marine and Atmospheric Science, performed the analyses. He charged less than a commercial laboratory and was very helpful. Gene and I had overlapped when he first came to the University of Miami Marine Lab. It was under what became his chemistry lab that I'd had my experience with the highly paid union plumber!

Using EPA-approved analytical methods, Gene found plasticizers and various pesticides. They were present in water from the well nearest the landfill, but he found even more in the well farthest from the landfill. The pollutants were restricted to freshwater in a highly porous zone above a regional impermeable layer just seventeen feet below the surface. Water below the layer was salty but free of pollutants! The bad stuff was close to the surface. We expected that the findings from our little fishing expedition would lead to a larger study. It seemed important to determine and map the total subsurface distribution of polluted water. We suspected the contaminated water was moving eastward and upwelling in Biscayne National Park. Doesn't that sound like something of societal importance that should be verified? We were about to learn yet another lesson about science and politics.

It wasn't long before the landfill manager, a Dade County employee, paid our little office a visit. He assured us that the landfill could not leak because it had an impermeable lining of lime mud! This time, he was the one doing the fishing! He was trying to find out who we were and what we were up to. He said the pesticides we found must have come from another source, possibly an older open dump down the road. He said he might come up with $25,000 if we did a study that would show the contamination came from the old dump site rather than his modern landfill. He also pointed out that the official EPA-mandated monitoring wells hadn't found any contamination.

We already knew those official monitoring wells were 100 to 150 feet deep. Someone had tipped us off. Our little effort had clearly shown that deep wells, even if directly under the landfill, could not possibly receive leachate. The regional unconformity and cap rock layer just seventeen feet below the surface prevented downward fluid movement. Besides, we knew there were even more unconformities below that one, and each one had a thin caliche cap. They were the same layers Ron Perkins used to identify and map subsurface Pleistocene units when he was part of Shell University, and they were the same layers that Ed Hoffmeister and Gray Multer had documented throughout the Florida Keys.

We had discovered that all the nasty liquids were restricted to a highly permeable zone, a porous one-foot-thick layer above the first impermeable layer. That permeable zone was laced with interconnected, tubular, finger-size holes made by burrowing crustaceans thousands of years ago before the sediment was converted to rock. When leachate-contaminated freshwater reached this zone, it flowed sideways. It could not penetrate downward through the impermeable layer and was forced to flow toward Biscayne Bay.

That permeable zone above the unconformity was destined to become very important to South Florida's drinking-water supply. Later, after the turn of the century, Kevin Cunningham and his colleagues at the USGS

Water Resources Division office in Fort Lauderdale would describe and determine its origin. His experiment with a red-dye tracer in that zone caused more than red faces at the USGS. The dye quickly reached the Dade County water-treatment plant, turning potable water red—so red that it stained thousands of people's underwear! But that's another story.

Yes, we could do a lot with our little drill, and it didn't require much money. In those days, our salaries were automatic. In the near future, USGS would change its financial procedures and it would become necessary to pay salaries from any outside grant monies. That change would put a damper on small exploratory efforts and thus the possibility of new discoveries. The little groundwater-pollution study we did with Sea Grant funding would not have been possible under the procedures that came later.

The new ruling that would take place would put the brakes on most venturesome exploratory research. Grants that would also pay salaries would have to be larger than most grantors were willing to pay. To make matters worse, there would be overhead costs, so what little was left over was often insufficient to complete a project. For example, it would have taken about $250,000 for three of us to initiate and officially complete the Mount Trashmore project (assuming we could get a permit). We made what we considered an important discovery with a thousand dollars.

Results of our little Mount Trashmore investigation convinced us we were really on to something big—something that had societally relevant implications, such as drinking-water contamination, groundwater pollution, and polluted water flowing below ground into a national park. Unfortunately, the mountain of garbage was already there and was not about to be taken down. In fact, it would double in size after Hurricane Andrew. In anticipation of a larger study on the underground leachate plume, we applied for a permit to extend our monitoring wells into Biscayne Bay, as we would then be working inside Biscayne National Park.

The permit request was rejected, and within a few months the Fisher Island Station officially and "coincidentally" came to an end. Like it or not, we were being transferred. After fifteen years at Fisher Island, we would all be gone by September 1989. I never for one minute believed there was a connection between our drilling, the pollutants we found, and our transfer, but Harold Hudson remained convinced there was.

Even after the transfer, we again tried for funding to continue the study. We didn't try for USGS funding. This time we teamed up with a University of South Florida researcher, because government agencies are not supposed to compete with universities for Sea Grant funding. That funding is for university researchers, but government employees could assist and be part of a university proposal.

The proposal was poorly written. In addition, at that time we didn't fully appreciate that this was the kind of work usually done by commercial consulting companies. Consulting companies contract directly with landfill operators. At that time I hadn't yet learned, and didn't want to believe, that consulting companies generally give the client what they want—a clean bill of health, or at least a modeling study based on little data that could not be understood by regulators or politicians. It was another lesson in the politics of science. We never followed up with this one. It was clear we would never be funded for this study no matter how hard we tried. We would eventually get back into the groundwater business, but this time we would be doing something the public wanted.

Some years later, I had the opportunity to accompany the then Secretary of Interior Bruce Babbitt, his science adviser, and an entourage of Park Service employees on a photo-op snorkeling trip in Biscayne National Park. As we stood at the base of the small coral rock lighthouse on Boca Chica Key, we could see Mount Trashmore looming over the horizon, about eight miles across the bay. My presentation was about the geology of the area, but that prominent mountain standing above the flat horizon was difficult to ignore. I decided to go for it and tell my Mount Trashmore story! I then recounted to the assembled group what we had found in our monitoring wells. The new park manager did not seem concerned, so I kept going. I soon learned that the science adviser to Secretary Babbitt had formerly worked for Waste Management Incorporated. He knew even more about landfills than I. When I finished, he said to me, "That's the story of landfills."

The one happy result of our work was that subsequent cells, the name for the various sections of landfills, were lined with a thick rubber lining. The first cell, the one around which we drilled, was underlain by a thin layer of lime mud. Before long, everyone knew that the first cell of Mount Trashmore was leaking. We did prepare a small report of our findings for the Park Service annual report, and it eventually became available on the Internet. Change doesn't come easily. It was yet another revelation about how politics and science interact. New discoveries with negative policy implications are not easily accepted. Ask any experienced Earth scientist.

Another useful outcome of our little thousand-dollar study was that a moat was later dug around Mount Trashmore. The moat was dug down to the top of the impermeable layer, and it quickly filled with the noxious fluid we had found in our first well. After I moved to St. Petersburg, landowners a few miles downwind of Mount Trashmore tried to bring me back to testify at public hearings. I resisted being drawn into the fray. Public hearings are not fun! You could be on the receiving end of a rotten egg or tomato, not to mention very unsavory words or problems with your employer.

Much later, I learned of a plan to pump the noxious moat water down one of the twenty-five county-owned deep-disposal wells located nearby. By then, Dade County was already pumping down about two hundred million gallons of treated sewage each day. The department operating the wells apparently told the landfill operators to drill their own well. I never knew the final outcome, but years later I learned that about ten of the twenty-five injection wells at South Dade had not been grouted properly and could not be used. They leaked, and contaminated water was migrating upward toward the freshwater aquifer.

Dr. Donald McNeill at the Rosenstiel School documented the leaking grout problem as part of his study funded by the Sierra Club. It wasn't the kind of study that leads to university tenure, as Don soon found out.

With South Florida's increasing population, it is only a matter of time before deep-disposal wells become a serious problem. No one really knows the full extent and capacity of the artesian Floridan aquifer. Where does the stuff go? The aquifer, generally known as the boulder zone, requires huge pump capacity. If the wells are opened, artesian pressure drives the water about forty feet above sea level. The cavernous limestone and dolomite aquifer is about twenty-eight hundred feet below the surface in South Miami. Something will eventually happen—I like to call it a train wreck—if we continue to pump fluids down there. Because of its depth and cavernous nature, the fluids may eventually well up from the bottom in the Straits of Florida where the boulder zone is exposed to the overlying Gulf Stream!

Ship Destroys Florida's Reefs

While working at Enewetak, we saw on the front page of a small island newspaper an aerial photograph of a grounded cargo ship. The headline screamed, "Ship Destroys Florida's Coral Reefs." After working in half-mile-wide hydrogen-bomb craters in reefs at Enewetak, the story and headline seemed insignificant. In fact, we joked about it.

Upon return to Florida, we learned that a ship had indeed smashed into a portion of Molasses Reef. The NOAA Marine Sanctuary management had brought a twenty-two-million-dollar lawsuit against the shipping company. To make the lawsuit stick, detailed mapping of the damaged area would be necessary. The offending ship was named the *Wellwood*.

Of course the grounded ship had not destroyed Florida's reefs, but it had indeed leveled more than two acres of Molasses Reef, one of the stellar attractions in the growing Keys Marine Sanctuary system. Harold Hudson offered his services and began making detailed measurements to legally document the extent of damage. He also transplanted huge hundred-year-old overturned coral heads, so they would not die or be rolled around by ensuing hurricanes. His work for NOAA brought in not only needed funding but also great appreciation. The study helped him make a career-changing decision. He had found his fork in the road and would remain in Miami to work for the Sanctuary program. Harold would not be transferring to our new base of operations in St. Petersburg.

Harold made the switch to the Key Largo National Marine Sanctuary, and in a few years the Sanctuary would expand to become the Florida Keys National Marine Sanctuary. As a result of his careful documentation, the shipline's insurance awarded the Sanctuary a settlement of six million dollars. The money was to be used for restoration purposes. By then, Harold, soon to be known as the "Reef Doctor," had found a home in NOAA, and that fine money kept him in the reef-restoration business throughout the Keys until he retired in 2007, with fifty years of government service.

Before moving the Fisher Island office to St. Petersburg, Barbara Lidz, Jack Kindinger, our new technician Frank (Pete) Spicer, and I prepared T-shirts that said, "St. Petersburg Here We Come." We probably should have said, "Look Out St. Petersburg!"

6

St. Petersburg and a New Beginning

Another Move

It was an emotional move. Once again, the Shinns were selling a home and searching for a new one, but this time they were out of practice. Fifteen years in one home was a new record for the Shinn family, and one can accumulate a mountain of difficult-to-move stuff during that time. Not to mention that we still had that five-hundred-pound cannon in the front yard! Moving was the bad news, but there was some good news. There were many homes on the market in St. Petersburg, and prices were way lower than in Miami. We nevertheless had some initial mixed feelings; Pat and I hadn't seen St. Petersburg since our high school band trip there in the early 1950s. St. Petersburg had become a very popular retirement area in the 1950s, and now retirees were expiring at a rapid rate. It was all about demographics. One joke was that the major exports from St. Pete were dead bodies and cut flowers. A newspaper report stated that St. Pete was the U.S. prune juice consumption capital. I heard an entertainer in the Keys say that "they have a drink in St. Pete called the Pile Driver. It consists of prune juice and vodka." Some rogues even called the city "the land of the living dead." It turned out to be not that bad, probably because we were rapidly joining the older crowd and developing a taste for prune juice!

We did very well selling our Miami home. In fifteen years our South Miami abode had quadrupled in value, and for the first time we could afford a home on the water—a canal connected to Tampa Bay and the Gulf of Mexico. The house we found had two stories, large trees, a swimming pool, and davits for our outboard boat. It also had a nice dock. A few years later I would be parking *Papa-San*, a forty-two-foot Kady Krogen trawler, behind our home. We bought *Papa-San* in 1996 with the money from selling my parents' home in Port St. Lucie after their deaths. I named the boat *Papa-San* and its support dingy *Mama-San*. Living on the water and owning a live-aboard boat had been our lifelong dream.

We moved into the new home on Labor Day 1989. Boy! Was it hot and humid! All we had heard about the heat and humidity was true. The west coast of Florida remains unbearably hot and humid during summer months. Miamians, including Pat, who was born in Miami, liked to call the west coast the "Backside of Florida." The cooling trade winds that bathe the east coast are absent on the backside. The summer wind that blows westerly from the Gulf of Mexico is very different and much warmer and more humid than the easterly trades.

The office space, thanks to business leaders and the University of South Florida (USF), was in a renovated two-story, red-brick building that had been a Studebaker automobile sales office that first opened in 1925. Red brick was another sign we were no longer on the east coast. The building was on the Historical Register and had been vacant for several years. It was home to rats, pigeons, mice, and cockroaches—excuse me, I mean palmetto bugs. It also served as a loft for local artists and various street people. After renovation, the ground floor, once the garage and showroom, became a machine shop and conference room. The second floor was converted into individual offices. A modern laboratory and new machine shop area were constructed several years later. Because the new building was connected to the building on the Historical Register, by law, it had to be constructed in the same red-brick style.

The business community, starting at the top with the mayor, as well as USF, was exceptionally receptive. The *St. Petersburg Times* wrote articles about USGS geologists even before we arrived, and most everyone in the city knew when we arrived. They really put out the welcome mat! That was a change from Miami!

The two-year undergraduate campus next door was enlarging, and the adjacent graduate school and marine laboratory were growing. The latter would become the USF College of Marine Science, and the two-year school became a four-year school. Over time I watched it all happen, not knowing that eventually the College of Marine Science would be my next move.

With growth and time came demographic changes. The population of St. Petersburg was steadily becoming younger. Unfortunately, I was not.

What was most difficult about the move was the sudden increase in non-research-related paperwork. It's called government bureaucracy. Fifteen years of running a four- to five-person research lab had not prepared me for the onslaught of non-research, often 100 percent time-wasting duties. The Fisher Island Skunk Works had been pretty much lost in the system, and communication was by mail and telephone. Eventually, we did acquire a fax machine that would send a one-page letter in six minutes! We thought that was a miracle! Those halcyon years were before computers and the fast-moving pace of the Internet. Now I had to shape up and join the computer age. The computer was great, but this advancement had a downside. Computers made it too easy for bureaucrats to request ever more paperwork, mandatory participation in online time-wasting courses that I used to call "classes in how to undo tangled paper clips," and enlarge their sphere of influence and importance. In no time at all, everything became BCOB (by close of business).

What made it all tolerable was my new boss, Bob Halley—the same Bob Halley who had started his career with us back on Fisher Island. Bob had left Fisher Island after six years and moved to Denver. After that, he became branch chief of the Woods Hole USGS office with more than one hundred employees to worry over. While he was there he had arranged the transfer of the Fisher Island feds into the Marine Geology Branch. By doing so he had saved us from being sent to headquarters in Reston, Virginia. Reston was close to Washington, D.C., the absolute epicenter of bureaucracy. In the meantime, Bob had learned the system, and besides, he was the natural team player that I never was. He grew up with brothers; I was an only child!

Why St. Petersburg?

In the late 1980s, the Woods Hole Oceanographic Institute, whose building housed the Woods Hole USGS office, decided they wanted their building back. They were disgruntled because USGS was not engaging enough Woods Hole researchers in their projects. When additional money was allocated, the USGS Woods Hole group would simply hire more personnel rather than share research projects with Woods Hole scientists. The Woods Hole office also had a rather elitist reputation within the USGS. That was probably because they were affiliated with a world-famous, highly regarded oceanographic institution. USGS scientists tended to identify with the institute's well-known scientists and to rest on their laurels. Within USGS, some people called the group Sleepy Hollow, contrasting it with the Menlo Park, California, office, which was much more energetic and aggressive and was sometimes called Hollywood.

Faced with eviction, the Woods Hole office needed a new home. Bob Halley, their branch chief, began searching. Many universities wanted them, but it was USF's Marine Science Laboratory and an influential group of businessmen movers and shakers called the Downtown Partnership who pulled the right strings to make the move happen.

They had an old building near downtown St. Petersburg that had once been the Studebaker dealership but had long been vacant. The building was condemned by the city and then bought for a song so it could be renovated and rented to the USGS. The old building had recently been put on the Historical Register. Why? The story I hear was that the man who owned it did that in retaliation against the city and politicians who had arranged for it to be condemned, thus reducing its value and forcing its sale to the university.

Bob's duty was to make the arrangements and select researchers from the Woods Hole office who would make the move southward. Not an easy task when dealing with dyed-in-the-wool Yankees in the Boston area. To us southerners, they represented the "stiff upper lip" for which many New Englanders are famous. Some in Woods Hole thought it was all a scheme to get rid of aging researchers and save money. It was thought they would not make the move south. And resist they did.

By the time the move was completely negotiated, an influential Massachusetts senator leaned on the Woods Hole Oceanographic leadership and put a stop to the expulsion. I suspect they were threatened with losing congressional earmarks and the like. Soon after that, the director of the Woods Hole Institute, who had put everything in motion, retired. The result of the fuss was that most Woods Hole researchers were allowed to stay, and only a small group made the move to St. Pete. I am grateful the group was small. There had been the distinct possibility that I would be forced to become one of the many geologists stationed at USGS headquarters in Reston.

This was another one of those major forks in the road. Eventually, Bob pulled it all together with the aid of Peter Betzer, director of the USF Marine Science Laboratory. Bob also had help from one of Betzer's former student colleagues, Bonnie McGregor. Bonnie had spent her first year in the USGS with our group at Fisher Island. She had worked for the NOAA Atlantic Meteorological and Oceanographic Lab in Miami before being hired by the Woods Hole group, and because her husband had a year of commitment to the Miami NOAA office, she couldn't move right away. We arranged that she would occupy an extra office with us at Fisher Island. Bonnie was a great addition and taught us a lot about continental-shelf geology. After a year with us, she moved not to Woods Hole, as had been expected, but to headquarters in Reston, where she became director of the Marine Geology Branch. She reported directly to the USGS director. Her husband, Bill Stubblefield, reported to NOAA headquarters in Silver Spring, Maryland, and later became an admiral in the NOAA Corps.

With both Bonnie and Peter Betzer pulling the right strings, the St. Petersburg office became an even more viable option. The red-brick vintage of the Studebaker building gave it a rather imposing look. The building reminded us of the Taj Mahal. Bob Halley was in charge, so some Woods Hole scientists who did not make the dreaded move southward began calling it the Taj MaHalley.

What was discouraging about the move for us was the lack of research funding. We Fisher Island feds felt like a little band of Skunk Works orphans. We had brought with us boats, drills, scuba tanks, compressors, microscopes, ship-to-shore radios, and other equipment, but there was no funding to get us in the field, and we were very field oriented. At the time, the new office was geared entirely toward the study of beach erosion and projects in the wetlands of Louisiana. What was that all about?

Wealthy homeowners' houses were falling into the water as beaches eroded away. They called their congressmen, and congressmen called USGS headquarters, and add-ons and earmarks would appear on congressional bills to fund beach-erosion research. These issues were the driving force behind research at the fledgling new USGS outpost. It was yet another example of how science and politics often blend together. Or as we would say, "We were meeting the needs of the public." It was "societally relevant research," a phrase we and other scientists had begun to hear increasingly more often—especially from those who wondered why we had been studying things like whitings!

I began to worry about what might happen when Congress eventually figured out that we could not stop the sea from rising. Sea-level rise, of course, was the basic driving force behind beach erosion, along with dredging of new inlets up and down our coasts. Inlets cut off the normal supply and transport of beach sand along the coast, and Louisiana was losing barrier islands at a rapid rate—about the length of a football field each year! There was little interest in coral reefs or research that might help discover oil and gas. My little group of orphans felt left out in the cold. We were literally put in a back corner of the building, and visiting scientists on building tours never quite seemed to make it to our corner. We were getting a little paranoid because it was so predictable.

An additional reason the St. Petersburg location was appealing to the USGS was the University of South Florida. USF is a state university, and being a state entity made it easier to pass money back and forth—something we could not do with the University of Miami, which is a private school. Thus, a large chunk of the USGS funding went to researchers at the USF Marine Science Laboratory. They could lobby Congress for funds, something that is illegal for a government employee or an agency such as ours. Call it "mutual back-scratching"!

For me and the other Fisher Island orphans, to work in the field required collaboration with the university faculty, who received the funds from the USGS. In many cases, they had lobbied for the money that went to USGS in the first place. The system had been set up to encourage that kind of collaboration. It was based on lessons learned at Woods Hole, where they had not done enough financial back-scratching. No one wanted to make that mistake again. In fact, the system worked well mainly because of some great researchers at the USF Marine Science Laboratory, located just a five-minute walk from the USGS Studebaker building. This orphan quickly joined forces there with Al Hine, who had similar interests in coral reefs and carbonate sedimentation.

The first five years were not my best. For a while I wondered if I had taken the correct fork in the road. One way I could keep my sanity was to walk over and visit with Al. He always had a way of cheering me up. Remember, at Fisher Island I could call my own shots, much as I had done for most of my career with Shell. It's not easy keeping an only child happy! However, I did adapt and eventually was able to find my own outside-funding sources and the freedom that goes with it. Bringing in your own funding from other agencies is the key to scientific freedom and productivity in the new bureaucratic world we are building.

One major frustration was that the USGS "hierocracy" had little interest in coral reefs. So here was the organization's main coral reef guy, unfunded to do coral reef research, except for what I could do with my colleagues at USF. It was doubly frustrating because I could clearly see that environmental issues associated with declining coral reefs would soon be big—very big! The USGS would eventually be dragged into the reef field whether they wanted to be or not. I can now look back and say how right I was. However, it took another fifteen years for USGS management to see how on target I had been, and by then I had new research interests—groundwater and African dust.

Before I move ahead with this story, the reader should be aware of another emotional issue to be overcome. Coral reefs were not my only interest. We worked on coral reefs at Fisher Island in order to determine why they grow where they grow. Such information could help exploration geologists searching for oil reservoirs in ancient limestone. We were not core-drilling reefs just to solve the kinds of environmental problems associated with dying corals, even though we had obtained outside funding to do so. Dying reefs were just beginning to be an issue. That would come later.

As the reader knows, I was equally interested in tidal flats, having done pioneering research on them in the Bahamas and the Persian Gulf. That was research that had application for oil exploration as well as for other environmental issues. People told me that in some quarters I was known as "Mr. Tidal Flat." It seemed logical from the start that I would become involved in the giant tidal flats of Louisiana. That was not to be!

Wetlands: A Road Not Taken

There were large tidal flats in Louisiana, but with the changing times came changing terminology. Tidal flats were now called "wetlands." I thought it amusing at the time. They were still tidal flats to me! They were the same as the classic tidal flats in Holland—the Wadden Sea—and the Wash in England, which had all been studied decades earlier to help understand ancient tidal deposits in the geologic record.

It took me a while to realize that Congress wouldn't fund anyone to study tidal flats. Wetlands were a different matter. The name implies essential home to birds and other warm-blooded cuddly creatures. They are also considered nursery areas for important commercial fisheries. This was biology, not geology, as I knew it. Tidal flats as I knew them had been studied to learn the sedimentary processes and the signature clues that would help identify analogues in the geologic record, especially those that contained oil. As mentioned earlier, a tremendous amount of oil was being produced from ancient tidal flats. Wetlands exist because of the same sedimentary processes, but now they were being studied for entirely different reasons. They were biologically important, and they were eroding and disappearing along with wealthy taxpayers' homes and businesses.

I was a little miffed that I hadn't been asked to join in on the new wetland studies. It was especially vexing because I had lived in New Orleans and had fished and hunted in the same wetlands that were now the subject of investigations. I knew my way around. And, there was something even more ironic about the situation. The Louisiana Land and Exploration Company, a major timber and petroleum-exploration concern, owned the major part of the Louisiana wetlands that USGS researchers were studying. The CEO of the company was none other than one of my former bosses—the man responsible for my going to Shell's head office. He was also my fishing and hunting buddy, Leighton Steward.

Leighton had important political connections and clout. Ironically, I had been in Louisiana and gone duck hunting with Leighton on their wetlands, but those in charge of the wetland studies never invited me to be part of their team. Was it a turf issue? I don't know. Possibly, there was fear of collusion and politics. I thought at the time the USGS scientists involved just wanted to do what oceanographers do because those in charge were basically coastal oceanographers. Oceanographers approach geological research differently than geologists. They had been developing sophisticated remote measuring devices to provide data for their models. They knew my approach would have been geological and more field oriented. I would want to take cores to identify various sedimentary environments. It was the tried-and-true way to understand sedimentary history. I also knew that subsidence and sea-level rise were the root causes of erosion and disappearance of wetlands—as well as the disappearance of important people's homes and property.

On the other hand, I knew there was only a certain amount of money available and I just didn't fit into existing funding plans. Leighton, I'm sure, would have supported my approach, and he had the ear of important politicians and landowners. He could have lobbied congressmen for additional

USGS funding. Feeling shut out, I returned to work in the Florida Keys. As I said, the first five years was tough on me, but one thing is for sure. I was always flexible and resilient.

It was gratifying to see that eventually a more geological approach to understanding Louisiana wetlands began to evolve. I suppose it had to happen, if only because it's easier to explain geology to politicians, funding agencies, and the public than it is to explain formulas for fluid dynamics.

Jack Kindinger from Corpus Christi had replaced Bob Halley at Fisher Island, and he already had experience with Gulf Coast geology. Jack rose to the needs and began conducting geologic analyses. His work involved shallow high-resolution seismic surveys that became known as "framework geology." He would eventually become a major player in the various wetlands projects and began taking cores and mapping the subsurface sediments. Jack would eventually rise in rank and become chief scientist of the St. Pete office. Meanwhile, I simply switched roads.

Back to Coral Reefs

Bob Halley, then St. Pete chief scientist, also wanted to work on coral reefs, but there was little interest in reefs within the organization at Reston headquarters. Coral reefs definitely were not where societally driven money was. Academics at the University of Miami experienced the same situation years earlier, even though they had coral reefs literally out their back door.

After about a year, Bob decided to step down as office chief at the St. Petersburg office. He'd had enough of holding people's hands, as he called it. He was dealing with a growing bureaucracy, complaining staff members, and dwindling funds all at the same time. I think most of all he was distressed because he too could not generate interest in coral reefs. Major funding was for tidal flats, aka wetlands, and beach erosion, and the master orchestrator and fund-raiser for those studies was Abby Sallenger. Bob decided it best that Abby take the office chief job. It made good sense. Abby was, after all, in charge of the largest chunk of research money in the office!

Because of the changing situation, Bob and I joined forces and set out to bring about change in our lives. We began by meeting with, and giving presentations to, citizen groups in the Florida Keys. They could help build support for coal reef research. Like us, they all knew reefs were dying. They didn't know why but agreed that research was needed. We also advised other agencies, namely, the NOAA National Marine Sanctuary, the Environmental Protection Agency, and the Florida Department of Environmental Protection, as well as Everglades National Park. We were critically aware that people in the Keys were concerned about their coral reefs. They called on us regularly.

What these citizens could do that we couldn't was talk to politicians and other agencies on our behalf. That was what beach-property owners had been doing for our beach-erosion and wetlands researchers. Together we cultivated many friends in the Florida Keys, where many already knew of my long history in coral reef research. They also knew I had practical knowledge of coral reefs gained from my years of diving and fishing—and they also knew I was a fellow conch! Life soon took a turn for the better and a new career opened for me, and for Pat as well.

Hurricane Andrew and Family Business

Three days after Hurricane Andrew struck in August 1992, I found myself on a USGS-chartered helicopter inspecting Miami from the air. The first day after the storm, I had been inspecting the Keys as a guest flying on a rich former-developer-turned-environmentalist's private airplane. No paperwork required for that one. This fellow knew every politician in the Florida Keys as well as those in our state capitol. He had also helped Bob and me become involved with the coral reef and Florida Bay problems in the Keys.

Effects of hurricanes on sedimentation and coral growth had been of great interest since my early Shell days. Mahlon Ball and I had described and published the geologic effects of Hurricane Donna, which crossed the reef tract in 1960. The Andrew storm was of particular interest to me. The hasty mission of these aerial investigations was to examine and document storm effects on beaches and to make photographic records. Although there was great destruction of mangrove shorelines, beach alteration turned out to be minimal because the powerful storm was small and impact was south of the major Florida beaches.

The tragedy was the effects of the storm on boats and coastal housing. I had been in Miami just two days before the hurricane struck, recovering our agency boat *Halimeda* and equipment—including that precious well-traveled underwater drill. We had just finished installing monitoring wells

In 1960, Hurricane Donna devastated coral reefs off Key Largo and overturned large elkhorn coral colonies such as this one. Gene's before and after photographs were pivotal to a classic paper by Mahlon Ball on geological effects of hurricanes.

Gene swims over the edge of a "blowout." Such sand-filled holes were made by Hurricane Donna and later by Hurricane Andrew.

for a major groundwater study. A few days after the aerial inspection, we were back in Miami with the *Halimeda*. This time we were making water-level measurements for the Federal Emergency Management Agency. We were determining the height of the storm surge mainly for insurance purposes.

With *Halimeda* we cruised Biscayne Bay from one end to the other and entered many partially destroyed homes along Biscayne Bay shores to record high-water marks. I learned that the best place to look for accurate high-water marks is in closets. We had the run of Biscayne Bay, which was littered with sunken and partially sunken boats. There were no enforcement authorities visible anywhere, as the entire government infrastructure of Miami was paralyzed and shut down. Cynical locals said it was because there was no air-conditioning in government offices. It was indeed hot and humid!

Needless to say, destruction was horrendous. Large boats had been washed ashore and stranded several hundred feet inland. Beautiful yachts were impaled on pilings. The neighborhood where I had lived for fifteen years was difficult to find. In many cases familiar landmarks were gone. My old home was still intact, but thousands upon thousands of homes in the area were in shambles. Many had been reduced to rubble.

With this storm came some personal serendipity, as anyone in the home renovation and window-treatment business could do well. Patricia had been at first devastated by the move to St. Petersburg. She was a dyed-in-the-wool Miami native, and she had had a successful window-treatment business in Miami, where she was known as "The Lady Draper." She had to abandon all that for the move to the "prune juice capital" on the backside of Florida.

By then our three sons were also in the window-treatment business. They had moved to Stuart, Florida. Pat had trained them all, including many of their friends. For the first year or so after our move she busied herself rebuilding and remodeling our new home, but that was not to last. She had to be active! She soon made contact with a well-known interior designer, and they did some jobs together in St. Pete. Pat had numerous contacts for obtaining window-treatment materials, namely, vertical blinds. Because of this, one of her first jobs was refurbishing the condo of a leading St. Petersburg Yacht Club officer, who was also one of the leading bankers in the area.

Then Hurricane Andrew hit South Miami and Homestead. The businesses of her friends and former competitors were in shambles, and her former customers were in need of replacement blinds and drapes—at least those who still had windows. Pat saw an opportunity and moved with great speed into her business mode. She worked out of our new home in St. Pete and made the five-hour drive to Miami at least twice a week. Sometimes she made the trip in three hours—and no speeding tickets! When her former customers called, they did not even realize they were talking to her in St. Petersburg, as she would often show up at their house in Miami the same day. Business was booming again! When she stayed overnight, she stayed on Captain Roy's boat, *Captain's Lady*.

Our sons were also doing well, but most of the manufacturers of blinds materials were in Miami or out of state. Our middle son, Tom, who by then had a wholesale blind-manufacturing business in Stuart, suggested that he and Pat set up a wholesale-manufacturing plant "in a more centralized part of Florida"—like St. Pete. They went into business together with Pat in charge of west coast operations while Tom continued to expand on the east coast. Business grew. Gene Jr. was doing the same but stayed on the east coast. Pat and Tom's business outgrew two different rented warehouses, and both finally moved into a much larger facility. Our youngest son, Dennis, became the main driver delivering the goods to retailers around the state.

In the late 1990s, the new fad in home decoration was plantation shutters mounted inside windows. Everyone wanted them, and whereas vertical blinds were cheap and highly competitive, shutters were new and more profitable. A larger workforce was required to meet demand. The company they created was "SHUTTER SMART," a division of South Florida Vertical Supply, which later became Southern Window Fashions. Soon there were five delivery trucks supplying customers over the entire state, and the business blossomed.

Another Bahamian Adventure

Hurricane Andrew, besides providing Pat a reason to get back into the window-treatment business, also gave me an opportunity to solve a mystery. For several years, Bob Dill and other geologists had puzzled over the layers of lime mud interlayered with oolitic sand in current-swept tidal channels in the Bahamas. We had seen the same things in Pleisto-cene oolitic deposits around Miami. How could this happen? Mud is a quiet-water, low-energy accumulation, whereas ooid sand requires fast-moving, high-energy tidal currents to form. Mud and sand just do not go together. Sand, regardless of its composition, is usually heaped into ripples and large submarine dunes.

Bob and I had long puzzled over this unusual relation. We had seen it in tidal channels around the Exuma Islands on many occasions, especially back when we were working on the stromatolites. We concluded that the mud had to be deposited during storms because of the objects we found embedded within it—but how? We found leaves and twigs in the mud, and once we found a three-foot-high Bahamian termite nest. The mud itself was laminated. Burrowing critters usually destroyed layers in mud because mud accumulates slowly, so there would be plenty of time for worms and shrimp to mix it up.

Because of our observations, Bob and I believed mud layers had to be some kind of storm accumulation. We envisioned a fast-moving mud slurry that somehow managed to settle to the bottom, but we had no proof. Then along came Hurricane Andrew, which provided just the opportunity we needed. To test our hypothesis we talked Captain Roy into using some unused charter time, so he took us, my wife, and Rick Major, a geologist from the University of Mississippi, to the Bahamas to investigate. No mess-ing around with the State Department for permits—we had to get over there while the evidence was fresh.

Sure enough, we found fresh layers of mud in tidal channels that were usually carpeted with rippled ooid sand. Mud layers were sandwiched between sand ripples, and they even contained fresh, green, sea-grass blades. The recent storm had indeed transported the debris into the chan-nels in the form of a thick slurry of mud and water. Problem solved! We published our observations, and geologists soon began recognizing simi-lar seemingly anomalous deposits in ancient limestones. Our observa-tions served as a guide to explain how the mud layers had formed—uni-formitarianism and serendipity had once again conspired in our favor! An untimely, devastating storm had provided an opportunity to advance knowledge about carbonate sedimentation. An earlier storm, Hurricane Donna, had already advanced our knowledge of hurricane effects on coral reefs and landward storm transport of sediments back in 1960.

Poop and Politics

The human population of the Keys was exploding at the same time reef corals were dying. The consensus of many was that coral mortality was caused by the increase in sewage—especially during the human surge called the "tourist season." Snowbirds—as Floridians call New Yorkers and other Yankees—flock to the Keys during winter months. At the same time, the area had become a NOAA National Marine Sanctuary to save coral reefs. Ironically, the publicity associated with the sanctuary also attracted increasing numbers of snowbirds and permanent residents, creating more poop. It was a self-perpetuating cycle, like the feedback that causes electronic amplification systems to screech when the microphone feeds the increasing sound to the loudspeakers.

The problem with blaming sewage as the cause of coral mortality was that it is very difficult to prove. Analyses of offshore waters were not pinpointing the sources or, for that matter, the causes of coral death. Many biologists were attacking the issue but were not providing satisfying answers. During my many trips to the Keys, I began to learn more and more about waste-disposal issues. I hadn't been aware that in addition to septic-tank systems with drain fields there were many so-called sewage-treatment package plants.

Restaurants and motels have what are known as "package plants." Package plants are simply small sewage-treatment plants that inject treated sewage fluids into shallow-disposal wells. The porous limestone of the Keys readily gobbles it up—out of sight, out of mind! There were at least a thousand disposal wells in the Keys. The lumps, so to speak, were removed, but the water that enters the subsurface still contains abundant nutrients. It is literally liquid fertilizer!

These gravity-fed disposal wells were originally only fifty to fifty-five feet deep. If you drill such a well anywhere in the Keys, the water level is just five to six feet below the surface, and water level in the wells fluctuates with the changing Atlantic tide—that's a good indication of how quickly water moves through the limestone. Tidal influence is especially noticeable in the upper Keys because those Keys are actually old fossil coral reefs and have very high permeability. Geologically, the lower Keys are very different. They are composed of oolite. They were formed by tidal currents and sand bars about 125,000 years ago. They are the same age as the upper Keys.

Some of the wells we drilled near Mount Trashmore, the giant landfill in South Miami, had already proven the presence of impermeable layers, some just inches thick. These thin, impermeable layers also capped several of the more deeply buried limestone units. The major one we had found seventeen feet below Mount Trashmore was a little deeper in the Keys. It ranged from about twenty to thirty-five feet below the surface. Just above it was the highly porous zone where the water rapidly moved sideways but not downward. Our work demonstrated that the direction of movement was toward the Atlantic—toward the coral reefs.

I began to suspect that if sewage waters were injected below the impermeable layer, the fluid might be trapped and possibly move laterally offshore. It might even go as far as the offshore coral reefs. Because the fluid was freshwater and lighter than saltwater, it might eventually rise up through cracks and breaks in the impermeable cap. Could it be that water was rising up and nourishing the ever-expanding acres of algae that were overtaking coral reefs? Well, it seemed like a plausible idea.

Algae, especially phytoplankton, which is microscopic algae in the water column, utilize the nutrients so fast that the amount in seawater is quickly depleted. That was why no one could find excessive nutrients in water over the offshore reefs. Could it be that nutrients from sewage were reaching the reefs from below? No one had tested this possibility. Once again, our little drill would be a key player.

The Florida Department of Environmental Protection—the agency that gives permits for package plants and disposal wells—had never conducted in-depth studies of nutrients in Keys groundwater. Population growth had been much too fast for them to keep up. Citizens groups wanted to know if sewage-package plants were causing reef demise and proliferating algal growth. We certainly agreed that it was a possibility. However, I did not point out that reefs all over the Caribbean in places where there was little sewage were also dying at the same time. That information would not have been well received.

Besides the thousand documented disposal wells in the Keys, there were also about thirty-five thousand septic tanks and about ten thousand so-called cesspits. Cesspits are simply septic tanks with their bottoms removed. They have no drain field like true septic systems. The rock is so porous that drain fields are not even needed. Where was the poop going? We thought we could find out. We just needed someone to pay for the research.

The USGS had little interest in coral reefs and there was little interest in a groundwater investigation that did not involve drinking water. Drinking water has long been a top concern of the USGS Water Resources Division.

Publicity, even bad publicity, can sometimes be a good thing. In this case it was a front-page news story about our work around a large sediment-filled sinkhole we had just found off Key Largo. The NOAA Undersea Research Program had funded the study that led to that discovery. The sensationalized front-page story caught the attention of the Florida DEP. The news story was also picked up and reproduced by the Associated Press and appeared all around the country, often with lurid headlines.

Some news stories said, "Sewage Killing Florida Reefs," says scientist blah-blah and so on. Angry dive-shop owners called me. It seems the comments attributed to me were hurting business. People were canceling dive trips! They didn't want to dive in sewage! I was terribly embarrassed, and from that experience I learned a lot about how the press operates. Newspeople who write stories do not write the headlines. Someone else does that. Newspaper owners, of course, are in the business of selling newspapers, so the more sensational the headline, the more papers they sell.

Not long after the news reports, a spokesman for the DEP called. He asked if we wanted to do a study for them. As it turned out, this fellow, Jack Myers, had also been a Shell geologist. In fact, the man who hired him for Shell was my friend Leighton Steward. Small world! We negotiated a one-year study, and I wrote a short proposal that was accepted by Carol Browner, who was then head of DEP in Tallahassee. Later she would be heading the U.S. EPA under President Clinton. After installing some twenty-five underwater monitoring wells, we had the project finished. We did it in one year, and publication of results opened new opportunities for study in the Keys.

That one-year study led to ten years of related research and launched what would become the "A-Team" into the world of hydrology, nutrient chemistry, and microbiology—all new to me! It also put me near the epicenter of Florida Keys politics, but I was squarely in my old stomping grounds. Finally, after all this progress, USGS headquarters management became interested!

With funding from the state and the federal EPA, I once again had freedom to call the shots, at least to some degree. With the funds, I was able to hire two great people. First was Chris Reich, who was graduating from nearby Eckerd College. He arrived at the office looking for work just as Pete Spicer, our technician from Fisher Island, was leaving to go back to school for a master's degree. Pete was headed off to Texas A&M.

Timing could not have been better! Chris and I, along with a summer scholarship student named Peter Cox, became expert underwater drillers and monitoring-well installers. It was interesting, hard work, and we were dirty and greasy most of the time. We became experts at installing shallow underwater ten- to sixty-foot-deep wells using scuba. Rather than simply drill holes, we also took cores. Now we also had cores from which to learn more about Keys geology as well as hydrology. And we did it all with Captain Roy!

We were drilling the last of twenty-five monitoring wells when we heard that a hurricane named Andrew might be headed our way. Captain Roy took no chances. He cranked up *Captain's Lady* and we headed to Miami towing *Halimeda*. As soon as we reached Miami, we dashed straight back to St. Petersburg. That night it became clear that Hurricane Andrew had become a major threat, and early the next morning—it was a Sunday—Pat and I drove the USGS vehicle back to Miami to recover *Halimeda* and all the research equipment still aboard *Captain's Lady*.

By then, Roy had already moved the boat to a protected area up the Miami River. We loaded everything we could and raced out of Miami before the major exodus began. We reached home in St. Pete after midnight. By early morning, South Miami and Homestead had been devastated, but Roy and the *Captain's Lady* were safe!

Creating the "A-Team"

During a side trip to Lee Stocking Island, where Bob Dill had discovered the giant living stromatolites, I met a young fellow named Don Hickey. Don was a student of Chris Kendall at the University of South Carolina. Kendall, of course, was the person who had worked in Abu Dhabi while I was in Qatar back in the mid-1960s.

Chris had been bragging about Don for some time, but there was a hiring freeze in effect so I couldn't hire anyone and there was no money. We were all on Lee Stocking Island core-drilling, but this time we used a drill developed by Dennis Hubbard of Fairleigh Dickinson University in St. Croix. Harold Hudson was also with us. We couldn't help noticing how hard Don worked and how quick and likable he was. Kendall had

The "A-Team" at work, 1995. Ann Tihansky records data while Chris Reich and Don Hickey operate a device for monitoring well pressure in Florida Bay. Photo by author.

Named after geologist T. W. Vaughan, this pontoon barge served as a platform for core drilling (and for sleeping) in the Florida Bay study of groundwater pollutants.

been right! We eventually figured out a way to bring Don on board in St. Petersburg through a contract company, and after several years he was finally brought on as a permanent USGS employee. Within a few years, he earned a master's in geology at the University of South Florida. The three of us (Chris Reich, Don, and me) began calling ourselves the "A-Team." A new person eventually came to our office and joined the A-Team. She was hydrologist Ann Tihansky.

The little family of A-Team drillers was thriving. We remained the A-Team for many years—and all this time we were still drilling with the aluminum tripod that Harold had fabricated on Fisher Island more than twenty years earlier! At this writing, it is still being used, and in 2010 Chris and Don with a dozen diving helpers core-drilled the Florida Middle Grounds in ninety feet of water. They found it had not been built by corals but by a worm-like snail that lives in the intertidal zone. It had stopped growing about nine thousand years ago when sea level was at least ninety feet lower than today. The A-Team lives on!

The one-year offshore groundwater study for the state and federal EPA resulted in a report six months later. We had learned a lot, and while doing so we had stumbled on a phenomenon new to us: "tidal pumping." Tidal pumping contributes to lateral groundwater movement in the Florida Keys and caused the rising and falling water we had observed in wells throughout the Keys. It was tidal pumping that made the water flow toward the coral reefs! We learned a great deal, but we could never prove that the sewage-contaminated groundwater was killing the reefs. I began looking elsewhere for a cause and came upon something totally unsuspected. The new hypothesis remains controversial, but I concluded that the demise of corals in this region was related to the increasing amount of African dust crossing the Atlantic Ocean. It led to a study that would continue until I retired in 2006.

Drilling the Everglades

The Everglades were in trouble. Some areas were dry when they should have been wet. Others remained wet all the time. Permanent water allowed exotic cattail plants to invade and overgrow the native saw grass, which is accustomed to periodic dry spells. Cattails first invaded the northern glades and began creeping southward. Water-management practices, sugarcane fields, the Corps of Engineers, South Florida Water Management District, and nutrients all took the blame. The public became alarmed and involved, and restoration issues began to evolve. Everglades National Park had sounded the warning bell and screamed, "The Everglades are not getting enough water!" They blamed the Army Corps as well as South Florida Water Management, and it wasn't long before the public became fully engaged.

Suddenly, USGS headquarters awakened! There was a water-related crisis brewing in South Florida and the Keys. But it was not the USGS Geologic Division (GD), for which we all worked, but instead the more visible and powerful Water Resources Division (WRD). They had long worked hand in glove with South Florida Water Management District in West Palm Beach. WRD was, and remains, the largest and most powerful entity within the USGS. They saw the need and the opportunity and quickly moved into the politically charged arena.

Fortunately for us, WRD could not ignore the two GD guys who had paved the way in the Florida Keys. Bob Halley and I already had our feet on the ground and had initiated groundwater research with funding from other agencies. We also had the drill, understood the local geology, the politics, the key players, and best of all, we could quickly and cheaply install monitoring wells. We could move fast, while WRD had to use contract drillers and all the paperwork and bids that go with contracting. The paperwork alone can take months. Fortunately, no companies were doing underwater drilling with divers so we came to a gentleman's agreement. "As long as we stayed offshore, there would be no conflicts," a WRD manager told me. The Everglades Restoration Project, or ERP as it became known, was a big-bucks operation. Officials, including Secretary of the Interior Bruce Babbitt, were talking about an eight-billion-dollar project to be spread out over twenty to thirty years.

Our groundwater research, initially funded by the federal EPA and state DEP, had finally morphed into a real USGS project with funding that came straight from Congress, albeit through WRD. Secretary Babbitt was squarely behind the Everglades Restoration Project. The result was that the A-Team finally received stable funding and we were on the move.

With increased funding, we began well-drilling operations in Florida Bay waters where the environment was in serious decline. Marine grasses and sponges were dying in the central part of the bay. Florida Bay was also part of, and administered by, Everglades National Park. The park administrators were fully behind our work. Although sports fishermen already knew there was a serious problem, Everglades National Park personnel quickly raised the Florida Bay situation to an even higher level of concern.

Interestingly, the people in the Keys who were initially concerned about coral reefs became unhappy. Attention that had been focused on the reefs was now being focused on Florida Bay, and coral reefs were taking a backseat. Nevertheless, both reef and bay lovers were united on one point—they both pointed the accusing finger at sugar plantations near Lake Okeechobee. The plantations and owners became known collectively as "Big Sugar." To the public, "Big Sugar" was aided by the South Florida Water Management District and the Army Corps of Engineers. Together the two agencies manipulated Everglades' water supplies to the benefit of sugar and other growers. The Army Corps of course had built the water-management system of canals, levies, and all the pumps that controlled the water supply. Their construction and digging activities had started after the flooding caused by a particularly wet hurricane in 1947. The Corps was very much involved, and conspiracy theories abounded!

It was not long before I found myself back in the mud—back where I started my Shell career. Besides drilling the limestone and installing monitoring wells, we began taking sediment cores of the numerous mud banks in Florida Bay. Pushing core tubes into the mud was one of my duties back at the Shell Development Lab in Coral Gables. It was, as Yogi Berra said, déjà vu all over again!

There was one difference. This time it was a large team, involving people from other USGS offices and agencies including South Florida Water Management District, and abundant money became available. We had chemists from our Woods Hole office and chemists from the NOAA Great Lakes Laboratory. Scientists smelled potential funding from miles away. My main contribution to this project was developing coring devices and transporting people safely to the inner reaches of Florida Bay for sampling and coring— and then back out. I was one of the few who knew the place like the back of my hand.

When not involved with the team effort taking push cores, the A-Team installed monitoring wells in the underlying limestone using a pontoon boat. We named our pontoon boat the R/V *Vaughan* after a famous limestone geologist who described the area in the early 1900s. Before long, we were venturing up the Shark River on the *Vaughan* and into the heart of the Everglades. On a few occasions we even stepped onto dry land to drill. Like the camel-and-the-tent fable, we were slowly working ourselves on-shore. By then WRD had accepted us on their turf, but only for drilling shallow wells. Once we even used Water Management District airboats and installed some wells in remote corners of "water-management areas" near Lake Okeechobee.

A-Team's Underwater Success

A major accomplishment for the A-Team was determining and verifying the direction of groundwater flow beneath the Florida Keys. We used a straightforward, labor-intensive method. No groundwater modeling for us; groundwater modeling does not work well in this kind of limestone and could not be trusted. Besides, we believed modeling was best used when the amount of tangible data is minimal—or sometimes nonexistent or unavailable. I remain suspicious of numerical modeling.

On both sides of Key Largo we installed circular arrays of monitoring wells. They were drilled into porous zones above and below the major impermeable unconformity. It was the same unconformity we had found a few years earlier at Mount Trashmore. The zone is a regional unit traditionally called the Fort Thompson Formation, but we knew the unconformity as the Perkins Q3 Unit. Two injection wells were installed in the center of the circular array, one above and one below the unconformity.

We injected two kinds of dye, fluorescein (yellow) in the shallow central well and rhodamine (red) in the deeper one. We could then sample the surrounding wells and wait for the first arrival of either dye. The time it took to reach a particular well gave us the direction and the speed of flow. Results? The flow was away from the bay and toward the Atlantic, and the fastest flow was in the shallow zone above the first unconformity. Groundwater was moving more than six feet each day. Tidal pumping was really working overtime!

We next installed wells in the dense wooded area of Key Largo, the island that separated the well array on the bay side from the well array on the Atlantic side. We could then trace the dye as it moved beneath the island and headed toward the Atlantic—and the coral reefs.

We also did some sneaky stuff. Remember the motto, "'Tis easier to seek forgiveness..." In the dark of night we spiked a major hotel's sewage-package plant with fluorescein dye. On another occasion I rented a waterside motel room and flushed the powerful dye down the toilet. That act gave us a bizarre idea, but we never acted on it. We were going to promote Florida Keys "Flush Day"! All we needed was a rich benefactor to provide packets of fluorescein dye to Keys residents. On a given signal over local radio, all residents would flush the dye down their toilets. We would then fly over and photograph the plumes of yellow dye emanating from the porous limestone. It would have been fun, instructive—and effective!

Another tracer we used was SF6, a clear, odorless, but powerful tracer. Colleagues at Florida State University provided the chemical and performed the SF6 analyses. In return, we let them use our various monitoring wells for their studies.

We made many friends by encouraging different groups to use our monitoring wells. There was only one problem—they had to find the wells. We made certain that there were no obvious markers. The wellheads were disguised as much as possible, and some even joked we disguised them for job security, but that wasn't the reason. Any unusual object on the bottom near the Florida Keys is quickly destroyed or removed by inquisitive divers and swimmers.

One question remained. What is the driving force that moves groundwater toward the Atlantic? We had suspicions. When drilling, we noticed that at certain times water gushed upward. A well could not be completed when that happened because water pushed the sand and cement out as fast as we put it in. We would have to wait until wells began sucking, which was the only time the sand and grout needed for well completion would go down. The sucking and blowing followed a six-hour cycle, so we knew it had something to do with the tides. The problem was there is virtually no lunar tide in the eastern part of Florida Bay. However, on the Atlantic side the tide fluctuates about three feet twice a day, especially during full moon and new moon periods.

During low tide in the Atlantic, we could attach a garden hose to our ocean-side wellheads and water would shoot as much as one foot above sea level! At the same time, wells on the bay side would suck water in. The reverse happened when the tide changed. Clearly, Atlantic tides were driving tidal pumping.

In a marine science class I took back at the University of Miami in the

1950s, I had learned that the water level in the Gulf of Mexico was actually about a foot higher than in the Atlantic Ocean. Using a surveyor's transit and instruments on wells on each side of Key Largo, Bob Halley proved this was still true. He determined water level in the bay to be, on average, four to six inches higher than on the ocean side. At certain times the difference was as much as twelve inches.

His measurements showed that some of the time, water level was actually higher in the Atlantic. However, during low tide in the Atlantic, water level in the bay was as much as three feet higher. It became a simple case of water flowing downhill. In this case, water was flowing downhill through the porous and permeable Keys limestone—toward the coral reefs.

Water would also move the other way for a short time when it was high tide in the Atlantic, but on average the net flow direction remained toward the Atlantic. Because of this tidal pumping, the flow would sweep up anything added to the groundwater under the Florida Keys—anything meaning sewage and whatever else went into septic tanks and disposal wells. "Flush Day" really would have been fun and instructive!

Everything we found made sense because we could watch the water flow in the tidal passes where there was no rock to slow the flow. We found this kind of flow everywhere in the Keys and later at Dry Tortugas as well. This was a major revelation for the A-Team, and we even won an award for our presentation about our results at the annual Society of Economic Paleontologists and Mineralogists convention.

Additional groundwater studies were also performed at the Florida Keys Marine Laboratory on Long Key. The lab, formerly a major tourist attraction called Shark World, had a small sewage-treatment package plant. We drilled a number of monitoring wells around the sewage-injection well and conducted numerous experiments. Chris Reich of the A-Team did his master's thesis research with dye experiments at this site. Many other people used these wells. Microbiologists Joan Rose and John Paul from the University of South Florida and their graduate students tested the use of harmless viral phages as tracers. Lee Kump and his students from Penn State used a safe radioactive tracer, and Jeff Chanton from Florida State University used SF6. In all the various experiments, the tracers soon showed up in the canal across the highway on the Atlantic side opposite the test wells. Later we did groundwater-tracer studies with Kump and students around a large

Gene often used a geological cross-section model of the Florida Keys to explain groundwater flow. This popular way of presenting science resulted in an elaborate ten-foot-long model constructed with the actual limestone that is presently on display at the Florida Fossil Coral Reef State Park at Windley Key in the Florida Keys.

treatment plant at Key Colony Village near Marathon, Florida. Again, we found that tracers moved toward the Atlantic and/or toward the nearest canal. Flow was never toward the Gulf of Mexico.

Lee Kump had worked with us at Fisher Island on a scholarship program before earning his PhD at the University of South Florida, so we worked well together. I cannot overemphasize the number of friends we made by collaborating with our wells and sharing the A-Team's drill.

Rose and Paul performed a simple but significant experiment at the headquarters of the National Underwater Research Program facility. This facility is located adjacent to an artificial canal with near vertical walls in Key Largo. They simply flushed fluorescein dye down the toilet. During the next low tide, about five hours later, dye from the septic-tank drain field emerged from the porous-limestone canal bank. It made a large yellow plume in the canal. Again, it was a case of water flowing downhill.

It became clear that what goes into toilets in the Florida Keys will reach the ocean and flow toward the Atlantic. How far it can go beneath the bottom and whether it emerges into the overlying water column farther offshore is still poorly known.

We had read about a device that was supposed to measure the rate at which water emerges from below the seafloor. Unfortunately, and after much effort, we found it does not work. Many researchers had been using what became known as seepage meters. There are many scientific papers based on the result of such meters. Most so-called seepage meters are simply cut-off ends of fifty-five-gallon oil drums. The drums are cut about a foot from the end, leaving a skirt of metal about one foot long. This skirt is pushed into the soft sediment to form a seal. A short pipe with a plastic bag attached is then screwed into the bung. Water moving up and into the space under the top begins to fill the plastic bag. The volume of water in the bag is then measured and a rate of seepage determined.

We knew from our drilling experiences that sediment, especially muddy sediment like that in Florida Bay, usually serves as a seal over the porous and permeable limestone underneath. In fine-grained sediment, mainly mud, water can only move up through burrow holes and around sea-grass roots. The reduced flow through mud means that flow is most likely to occur where there is bare rock and no sediment. Unfortunately, conventional seepage meters cannot be used on rock bottom. To measure the volume of groundwater emerging from the bottom, the A-Team devised another approach.

To get a better idea of flow rate directly from the limestone, we constructed fifty fiberglass domes, each with a port at the top to which we attached a plastic bag. Using quick-setting underwater cement, we sealed the domes directly to the bare limestone. The bags filled quickly. In fact, we measured such large volumes that we quickly calculated that all the water in Florida Bay could come from groundwater seepage and it would only take about six months. That seemed too good to be true, and yes, it was!

The A-Team finally performed some simple experiments that proved our data invalid. I was crushed! Finally, to really understand what was happening, we conducted a series of controlled experiments in my home swimming pool. For the published paper we called the pool a "test tank" ("swimming pool" simply would not be accepted in a scientific journal). We placed seepage meters in the test tank and identical devices in the shallow bayou about a thousand feet from my home. All the seepage meters were placed in five-foot-diameter sand-filled plastic pools made for children. We placed these sand-filled pools in both the test tank and bayou simultaneously. If the plastic bags filled, we knew it could not be groundwater, and certainly there was no groundwater in the test tank. Nevertheless, those in the bayou, where there were small waves, collected water.

In another test, we placed the plastic kiddie pools in Florida Bay next to our monitoring-well array. We proved that the water we collected was not groundwater. We and all the other scientists who had used seepage meters had been victims of the Bernoulli effect! That's the same force that gives airplane wings their lift. In this case, it was the back-and-forth movement of water caused by waves passing above that created the sucking action. In the test tank there were no waves and there was no flow. In the bayou there were both. It was as simple as that. It was a great blow to the A-Team, and it was also a blow to others who had relied on seepage-meter data. It was doubly disheartening when we found that bags would fill up even when adjacent to a monitoring well that was at the same time sucking water in!

We had wasted a lot of time and money in construction, installation, and measuring, not to mention the long, boring drives to and from the Florida Keys. To warn others using seepage meters—many published studies had depended on them—we published a paper titled "Seepage Meters and Bernoulli's Revenge." We took special pains to point out that seepage meters do indeed collect near-surface groundwater that can be used for chemical analysis, but the volume of water should not be used to calculate seepage rates. Seepage meters might work in places where there are no waves. Many researchers were unhappy with us, and I confess the results did not please the A-Team either. Nevertheless, we remain certain that groundwater is indeed emanating from the bottom. After all, our wells clearly showed that the pressure is high enough to force water above sea level. The problem is that seepage meters, as presently designed, tend to create artificial flow when there are waves or currents.

One triumph did come from our various studies—the state of Florida was quick to mandate that disposal wells be drilled deeper. In the beginning, legal depth was fifty-five feet; when we began our studies it was sixty-five feet. Because of our work, the state mandated they be deepened to ninety-five feet. The A-Team could feel proud that we had at least accomplished something with practical societal application. Here was an example where policy and regulations actually did result from scientific data rather than political or emotional whims.

A major Florida Keys problem remained. Why were the corals dying? Throughout our studies, coral diseases and proliferating algal growth continued unabated. The A-Team did not know why, and to complicate matters, the problem was not specific to Florida. The same coral death was being observed in the Bahamas and elsewhere in the Caribbean. I was about to be off on yet another fork in the road. Although it had been an uphill battle convincing geologists that sediment was turning into rock in the Persian Gulf—or that mud was precipitating to form whitings in the Bahamas—what came next was much more contentious.

Killer Dust! A New Major Adventure

In 1984, I attended a geological meeting and field trip on San Salvador, an island in the eastern Bahamas where Christopher Columbus had supposedly landed in the New World. The meeting was held at the Finger Lakes Marine Laboratory. My wife and I were there with Phil Dustan, who had come from his new job at the College of Charleston. For some time, Phil and I had been mystified by the rapid demise of corals, especially my favorites, the elkhorn and staghorn species.

San Salvador is the most remote island in the eastern Bahamas. It is surrounded by deep clear-blue water, and the human population is small. One day our guides took us to a reef within swimming distance of shore that had been the major attraction for a nearby resort that catered to underwater photographers. They called the reef "telephone-pole reef." In 1983 the scientists who managed the marine biological station had watched the reef die, and they told me the reef had died over a short period of time—in about one month. The dive resort consequently went out of business.

At about the same time, the common black long-spined sea urchins called *Diadema antillarum* began to die—not just in San Salvador, but also everywhere in the Caribbean. Progression of the disease and their death were well documented and had been published in the journal *Science*. Apparently, at the same time, a disease affected sea fans nearly everywhere from the Caribbean to Bermuda. Sea-fan disease was not so obvious and did not receive a lot of attention at the time. What we could all see was that so-called turf algae had proliferated nearly everywhere in the Caribbean.

The entire coral research community was aghast and in shock over these events. At first the public and sports divers paid little attention. They were unaware of the extent and severity of these events, which they thought were a local phenomenon. We had all observed increasing coral death in Florida, but in San Salvador we were in the far eastern Bahamas on an island away from everything—or so it seemed. Despite its small human population, corals were dying there just as in the Florida Keys. I remember looking at Phil in amazement when he said, "It's as if it's coming from the sky." Outrageous as it seemed, the comment stuck.

More than a year after the San Salvador experience, I read a short article in *Geotimes*, a small trade magazine for geologists now called *Earth* magazine. The article described how African dust crossing the Atlantic nurtures the Amazon rain forest, and it noted that rain forests are notoriously low in nutrients. A friend who operates research stations in the Amazon Basin later told me that electric fish-shocking devices wouldn't work in Amazon-basin streams. The water is salt and mineral free—almost like distilled water because it originates from rain and melting snow—so some salt has to be sprinkled in the water to make the shockers work.

I also learned of a unique community of air plants, called epiphytes, living in the Amazon tree canopy dozens of feet above flood level. The roots of these air plants are enshrouded with red soil, but they are way too high to be reached by floodwaters. Possibly termites could transport it there; nevertheless, the red soil is African soil! According to the short article, the dust delivers about half a pound of phosphate to each acre of forest every year. The Amazon forest actually benefits from nutrients in African dust transported three thousand miles across the Atlantic. I found that amazing!

Remembering what Phil had said, I wondered, "Could that same dust be fertilizing coral reefs and stimulating algal growth? Could it even explain the problems in Florida Bay where sea grasses were dying?" Maybe the problem in the bay was not nutrients from the sugar plantations after all but dust from Africa. What an outrageous idea! "Better not mention that one," I thought. "People will laugh." And they did!

The next surprise was learning about the amount of dust that crosses the Atlantic. That little article had stimulated me to learn more, so I dug into the published literature. I was further amazed. The amount of transported dust reaches into the hundreds of millions of tons. About one billion tons leaves Africa every year.

Geologists knew dust had been transported for thousands of years. Throughout the Florida Keys, there are layers of dense red/brown limestone capping porous limestone. The red/brown-colored cap is caliche. Some call it calcrete or soilstone crust. It was the impermeable layer we had found under Mount Trashmore! That work was done before we learned that the red color was from iron in the dust. When not cemented into rock, the dust also forms the red soil that supports agriculture in the Bahamas and Bermuda. In San Salvador, the red soil is called pineapple loam.

Now we knew why the layers on top of the Q units we had identified while drilling in the Keys were red to brown in color. The dust is mainly silica and clay minerals but contains about 6 percent iron, which gives it its color. Clearly the dust—red/brown because of its oxidized iron—has been blowing across the Atlantic for aeons. Has something changed? Why would it affect corals now and not before?

During the San Salvador trip I learned from an archaeologist that when Columbus landed he had found pottery made by indigenous people. Clay minerals are necessary to make pottery. It's the clay that hardens when clay pots are fired. There is no source of clay anywhere in the Bahamas or South Florida. Even if you drilled down several thousand feet in the Bahamas, there would be no river-borne clay. What little there is on and in the rocks comes from Africa, riding on the trade winds. Natives had simply scraped the clay from limestone surfaces, mixed it with water, shaped it, and cured the vessels with fire.

I kept searching the literature and learning more and more. Joseph Prospero at the University of Miami Marine Lab had begun monitoring dust on the island of Barbados back in 1965. I knew about his work but never understood why he did it. After some digging, I understood. Joe had originally set up the monitoring station not to catch African dust but to collect dust from outer space. What he got instead was red/brown soil particles. It could only have come from Africa. More investigation revealed that even Charles Darwin had reported African dust falling on the HMS *Beagle* during his famous worldwide cruise. He reported it in 1846 and noted that it caused mechanical problems and eye irritation among the sailors. Much later, it was recognized in sediment cores from the Atlantic.

A graduate student at the University of Miami named Nancy Maynard had identified African dust in Atlantic sediment cores. How did she do it? She found microscopic freshwater diatoms, tiny green algae that grow only in freshwater lakes. There were also curious silica bodies called phytoliths. Phytoliths come from various grasses on land. They are the "saw teeth" that give saw grass its name. So even before satellite images became available, researchers realized the red dust had come from Africa.

Initial research focused on the meteorological processes by which it can be lifted and blown so far. Those researchers were also determining how much dust was blowing out to sea and across the Atlantic. After Caribbean reefs began dying, my questions became, has it happened before, and does the dust do anything else besides provide material for pottery, soil for agriculture, interesting layers for geologists to study, and fertilizer to keep the Amazon rain forest healthy? I suspected it did a lot more!

What was exciting about the quantitative sampling that Prospero had doggedly conducted since 1965 was that the amount of dust transported fluctuates from year to year. There are good years and bad years, so to speak. Since 1965 there had been peak years in the early 1970s centered around 1973. But what a surprise when I learned that the biggest peaks of all had occurred in 1983 and 1984, the year telephone-pole reef had died. It was also the same time that sea fans had become diseased, and sea urchins had died throughout the Caribbean and in Florida!

This was exciting, but I really didn't know what to do with these isolated facts and observations. What might they ultimately mean? Soon there was an event that directed me down a very dusty fork in the road.

A Dusty Road

In 1996 the guest lecturer at the University of South Florida was Richard Barber from Duke University. He had done research for the American Petroleum Institute back when I worked for Shell, and we had met briefly at a meeting devoted to tar balls on the beaches of Bermuda. Barber's lecture at USF described an experiment in the Pacific in which he was part of a research team that discharged a large volume of liquid containing iron sulfate into the sea. Why? That part of the Pacific has very low primary productivity. There are few phytoplankton in the Pacific west of South America, and phytoplankton are at the base of the ocean food web. The waters where they experimented contain sufficient nitrogen and phosphate for plant growth, yet little growth is found in that region. Earlier work by John Martin, the director of Moss Landing Marine Laboratory in California, had indicated the missing nutrient was iron. He had worked long and hard to make this large experiment happen, but it required top-down support from key congressmen. Many scientists were against it. According to Barber, the experiment would have been rejected by a bottom-up approach. The bottom-up approach, the way most science is funded, requires peer review and consensus—in other words, scientific support from the bottom up. Controversial ideas often do not make it through this long and sometimes convoluted process.

The main message here is that the massive seeding experiment—code-named "Iron-X"—proved the validity of Martin's hypothesis once and for all. Within a day of discharge, a patch of clear-blue ocean turned green in a patch large enough to be seen from space. His lecture served as a trigger—an "aha" moment. I knew for certain—African dust is loaded with iron! So, after his lecture, I coaxed Richard to my office, told him my idea, and showed him the graph of Prospero's dust data and the correlation of dust peaks with coral demise. Richard was intrigued. He already was convinced that human sewage pollution could not explain simultaneous reef demise all over the Caribbean. Before he left he asked if I would be a coauthor on a poster he and others were presenting at the upcoming Oceanography Conference in Seattle. I agreed and then mentioned it to my colleague Terry Edgar.

Terry said, "Why don't you do a stand-alone poster yourself?" He knew I had been photographing and documenting the demise of corals over the years and had some striking serial underwater photographs starting as early as 1959. They showed when and how fast the corals were dying. Terry convinced me. Barber just happened to be on the Oceanographic Society's Steering Committee, so he pulled some strings and my poster presentation was arranged. While I was hesitant before, now I was committed. It was a major turning point.

The poster presentation was a success, but at that time I was thinking only of iron. I reasoned that if iron could make a portion of the blue Pacific green, maybe it could stimulate the growth of benthic algae on coral reefs. It seemed reasonable, but it still didn't explain all those coral diseases. Could iron stimulate the growth of bacteria? Some people thought so. Could the dust carry disease organisms?

It was about this time that I looked up the papers by Nancy Maynard. She had been interested in dust from her graduate school days but could never obtain research funding to follow it up. She had once deployed sheets of wet cheesecloth in the air to collect dust. She caught not only dust but also living microbes! Still, she could not obtain funding, so she took another road, a large one, and ended up on the White House science staff.

Before beginning her White House duties, Nancy had worked for a while on tar balls at Bermuda. She was also involved with the American Petroleum Institute, which funded Barber's research back then. I had known Nancy when she was a graduate student in Miami, but I'd never known of her interest in dust. After the tar-ball work she was drafted into the science staff of President Ronald Reagan. She stayed on and also served under the elder President Bush. After that she went to NASA and became head of Outreach and Applications.

I wrote Nancy and told her what I had been doing since we had last met. What timing! She just happened to be setting up a new research project at NASA Goddard Space Center on human health. She was especially interested in the worldwide increase in asthma. Naturally, she already knew that breathing dust causes respiratory problems. More important, she knew that asthma had been increasing throughout the Caribbean Islands. It had actually increased seventeenfold at Barbados since 1973. Asthma, in the sunny Caribbean? That was a surprise. But the date seemed to fit—1973 was the first spike on the Prospero dust graph.

We hit it off right away. She asked, "If I get you some money, what would you do with it?" I had an answer ready. I had served on the graduate committee of Dale Griffin, a student in public-health microbiology at the University of South Florida. Dale had also helped us with the groundwater work in the Keys. In a millisecond, I said, "I would hire a PhD microbiologist to see if there are live microbes in the dust. I just happen to know one, and he is looking for a job."

It wasn't much money, but it was enough to pay his starting salary. Like Don Hickey, we had to bring him aboard through a contract company. Dale moved quickly. We had already been discussing the possibilities of microbes in the dust for several months. I had even approached one of his advisers, a well-known microbiologist named John Paul who explained how NSF funding works. He had an established track record for what he was doing and told me the reviewers would eat my lunch if I proposed looking for microbes in transatlantic dust. That's an example of the problem with the bottom-up approach to funding that Richard Barber had informed me about! This was also the very problem in the science peer-review process that I had complained about for many years. Every scientist who depends on external proposal-driven funding knows how difficult it is to take on new and different research. Once you become established in a certain field, you can easily get stuck in a funding rut. Fortunately, Dale was just starting out on his new road.

About this time, I had made friends with a new USGS colleague in the Virgin Islands, Virginia Garrison, known to her friends as Ginger. She and her husband had operated a charter boat service in the Caribbean, had lived on a boat, and had worked for the National Park Service before being transferred into the USGS Biological Division. She had traveled all over the Caribbean and had toiled many hours cleaning the red dust off her boat. She knew all about African dust clouds and in addition was documenting the death of corals in the Caribbean. She jumped on board!

Ginger began her career as a chemist, so with her background she quickly devised a dust-sampling device that pulled dusty air through a sterile Millipore filter. She collected samples from the air when dust rained down on her boat and also at times when there was no visible dust. Almost immediately, she began mailing samples to Dale, who by then had developed a method for culturing microbes without removing them from the filter.

Dale, like most other microbiologists, had been taught that microbes do not survive long in the atmosphere where they are exposed to deadly UV rays and desiccation. Ultraviolet light, as many readers know, is often used to sterilize toilet seats! With this background—paradigms again—he was surprised when cultures began growing on the filters. Only a few grew on filters deployed when there was no dust, but filters that were set up during obvious dust events blossomed with bacterial cultures.

Within six months, Dale had cultured and identified sixty-five species and prepared his first manuscript for submission to the journal *Science*. His study found that about 10 percent of the microbes were known to affect humans with damaged immune systems (think AIDS). Thirty percent turned out to be plant pathogens!

Science rejected Dale's article. This was just the beginning of resistance to what became known as the "dust hypothesis." Dale later published the article in a specialized journal called *Aerobiologia* devoted to atmospheric phenomena. With the NASA money and Dale's published results, we had our needed jump-start. Soon USGS money became available, and eventually there would be more substantial funding from a congressional "add-on." We thank our Florida congressman Bill Young for that. Now that's the top-down approach! Dyed-in-the-wool bottom-up people might call it "Pork."

Again our relationship with the University of South Florida College of Marine Science paid off. It helped greatly that Dean Peter Betzer had done his dissertation on Asian dust effects in the North Pacific. He had shown that dust increased primary productivity. It's like fertilizer for phytoplankton. He also knew all about the Iron-X experiments. Peter had gone to bat for us with his friend Congressman Young.

Soon Chuck Holmes and his technician Marci Marot demonstrated the presence of various isotopes, such as beryllium-7 and lead-210 in the dust. One sample collected in the Azores after a large dust event was especially rich in beryllium-7. The gamma count was 43,000 disintegrations per minute (dpm)![1] Chuck and Marci had never seen such high levels. We also found high levels in the dust that accumulated in the bottom of drinking-water cisterns in the Virgin Islands. Sediment in one ten-year-old cistern had beryllium-7 levels of about 500 dpm! It was dropping out of the atmosphere along with lead-210, and not only that, toxic metals such as mercury and arsenic were also present.

Such high amounts of beryllium-7 are so rare that there were no established health-standard levels assigned to it. However, there are standards for lead-210, which also emits gamma radiation. Chuck noted that if the gamma radiation they measured in beryllium-7 had come from lead-210 (for which there are OSHA standards), it would have been three times that allowed in the workplace. So, this was what people were breathing when dust storms sweep through the Virgin Islands. No wonder they cough and wheeze, as did the sailors aboard the *Beagle* during Darwin's day! What else were they breathing? Live bacteria and fungi, radiogenic isotopes, mercury, and arsenic, and then there are the pesticides such as DDT that are still used in Africa to fight mosquitoes and locust plagues. Did this suggest a connection with the seventeenfold increase in asthma in the Caribbean? You bet!

One of the first fungi identified in the dust was a soil fungus called *Aspergillus sedowii*. Although soil fungi do not reproduce in seawater, it was this same fungus that caused the sea-fan disease. Garriet Smith, a microbiologist at the University of South Carolina at Aiken, had made the discovery of that fungus. He and others had found that the disease began throughout the Caribbean in 1983. Here was our smoking gun! It indicated that other diseases affecting corals and sea urchins in 1983 and 1984 might also be caused by something in the dust. These were heady times for our emerging group. We didn't know where it would lead, but we did know we were on the right road.

It wasn't long before we recognized the possibility that terrorists could theoretically use dust as a cover and a carrier for some nasty bugs, such as that described in the introduction to this book. Bioterrorism attacks seemed a very real possibility, and those thoughts kept me awake at night. What if someone put some really bad stuff—for example, the anthrax bug—in the dust before it left Africa? That really worried me. My friends, of course, thought I was being melodramatic or just plain crazy.

Brick Walls and Agency Agendas

It was an exciting research time for all of us, but we would soon run into un-expected barriers. We never knew what would turn up in the next sample. However, it soon became apparent that many decision makers—especially those controlling the purse strings—didn't really want to go much farther. After all, there wasn't much we could do to stop the movement of African dust across the Atlantic. It also became apparent that such research might undermine programs and agendas in other government agencies.

For example, the Department of Agriculture spends an enormous amount of time and money on agents who examine luggage for fruit and plants that might contain some bad bugs. They examine shoes and boots for soil that could contain bad bugs—stuff that can affect humans and plants might arrive in the luggage areas of jet planes and in boxes of imported foods. These are expensive and necessary activities to keep bacteria and in-sects out of the country. How would they deal with the knowledge that the millions of tons of soil blowing overhead might be transporting the same bad bugs? What about insects, various mites, and mosquitoes? These kinds of bugs had shown up in Cuba, and the Cuban government had blamed us!

In 1988, African locusts, some measuring two inches in length, had coasted on dust-laden trade winds, and countless millions landed on eastern Ca-ribbean Islands. The aerial invasion of these locusts was well documented and had been reported in the scientific literature. Their arrival explained the presence of more than a dozen species of African locusts already estab-lished in the eastern Caribbean and in South America. To our knowledge, the Department of Agriculture paid little attention. "Don't locusts eat crops?" we asked.

With each discovery, we became more and more excited. Since we began the research on dust in the late 1990s, numerous tiny plant-eating and sap-sucking insects of African origin have appeared in South Florida. As of this writing, Douglas Seba, my fellow dust researcher who lives in Key West, has just discovered that 0.5-millimeter-long animals called rotifers also are abundant in African dust. Rotifers, mainly the freshwater variety, produce microscopic eggs while the animals themselves shrivel up when the water evaporates. The eggs and shriveled and dried adults can survive at least nine years in this state. Once wet, the dried rotifers and eggs resume a normal swimming life within a few hours! It's no wonder that rotifers are worldwide in their distribution. They can blow anywhere surviving both heat and cold.

It was the invasion and outbreak of sugarcane rust and tobacco-killing pathogens that led Cuba to accuse the United States of bioterrorism in the first place. That they were not targets was indicated by the fact that surrounding islands in the Caribbean also experienced the same invasive species. At the time, few people knew that insects, microbes, and viruses could make the transatlantic trip without human assistance. Ironically, checking for fruit and dirt on travelers' shoes and luggage remains our major defense against such exotic invasions.

Another Brick Wall

From the beginning, especially after seeing the dead coral in San Salvador and discussing the problem with Richard Barber, I suspected something in the dust could be the cause. That concern is what in fact had started this whole line of investigation. Discussions with Richard and preparing my first poster had convinced me of a connection between dust and coral demise. The timing was right, because Caribbean-wide coral death occurred during the peak years of dust transport. However, correlation is not considered scientific proof of anything, though it is a good start toward obtaining more evidence. Unfortunately, there were many other factors that argued against the dust hypothesis.

What I failed to appreciate in the beginning was that agencies protecting coral didn't want to hear this hypothesis. Why? Well, they may be able to regulate sewage and other human activities, but they were powerless against soil crossing the Atlantic. If they agreed that soil was the cause of reef death, they might as well throw up their hands in defeat. Some might even lose their jobs! If they publicly agreed with the hypothesis, they might be accused of wasting taxpayers' money on needless enforcement. Although those agencies support research, there was little incentive for them to support this research even though it involved life and death of coral reefs. Might our results indicate that the reef managers had been on the wrong road? Again, "When you're on the wrong bus, every stop is the wrong stop."

In spite of these issues—and initial rejection by the journal *Science*—I finally published my first paper proposing that African dust was the cause of reef decline. It was published on October 1, 2000, in *Geophysical Research Letters*. Getting it accepted for publication, even with well-respected coauthors, was painful. Joe Prospero, Dr. Dust himself, was one of the coauthors. The effort reminded me of what we faced getting our paper on whitings published! In that case, it was only the egos of certain scientists at stake rather than entire agency policies.

It didn't take long to appreciate how our study might affect other agencies, especially the U.S. EPA. Even though Prospero had shown that African dust sometimes exceeds EPA particulate standards in Miami, the EPA didn't know how to handle the issue. Why? The EPA had been in litigation with industries and their chemical and particulate emissions for at least three decades. If the EPA funded our work, industry lawyers would have latched on to our data and used it as part of their defense. The EPA was well aware of our work because we had made several presentations to EPA researchers and managers.

Meanwhile, everywhere we gave presentations there was enthusiastic interest. It was impossible to give a presentation to a group that didn't contain people—or people with children—who suffered from asthma or other bronchial distress. The American Medical Association didn't jump on board, but one medical group was very interested.

The American Academy of Environmental Medicine quickly recognized the importance of our work. I was invited to give presentations at three of their annual meetings. Why were they so interested? The academy consists of physicians who treat people with allergies and susceptibility to various chemicals. The first meeting where I spoke included a group of ex-military nurses. They had served in the first Iraq war—Desert Storm. Some were in wheelchairs, most had visible tremors, and all were convinced the malaise was caused by something they had inhaled. I was nervous about the presentation. I had never addressed a group of medical people before. When I finished, some of the Desert Storm survivors hugged me. They all complained in private that the Veterans Administration had brushed them off. Apparently, I was the first one with data indicating that dust with all the unhealthy stuff it can contain might be the cause. A doctor from England also presented a paper at that meeting. She treated British military from that war and explained that they were dying at a rate of one a week. She mentioned that the British army medical people had lost most of their soldiers' medical records.

African Dust and Homeland Security

In Puerto Rico, I gave a poster and a presentation at the annual meeting of the Armed Forces Pathology Research Institute. Chip Groat, then director of the USGS, was interested in our work and had been encouraging USGS researchers to apply geological research to medicine and human health. That was the main reason I felt it important to present our findings to the Academy of Environmental Medicine. In his plenary address to the Armed Forces Pathology meeting, Groat mentioned our work and showed a now-famous satellite image of a dust storm crossing the Atlantic. Having the director present that image certainly captured everyone's interest, including that of an Air Force general.

A general from Fort Detrick heard the talk and later approached me in the hall. He began telling me how tough the anthrax bug is. He said, "The anthrax bug would be here in a few days if someone spread it on the desert in North Africa." Next he said, "Do you need funding? Do you need aircraft? I can get both for you." I was excited and began reading all I could about weaponized anthrax. I read Russian defector Ken Alibek's book *Biohazard*, where I learned that during the Cold War the Russians had produced thousands of tons of the stuff. Where was it all? I read Judith Miller's book *Germs* and Tom Mangold's book *Plague Wars*. It became difficult to sleep after reading those tomes. Meanwhile, I waited to hear from the general.

Six months went by and still no word. I wrote a letter—no response. Finally, when we had a few key publications under our belt, I wrote again.

This time it was a two-pager outlining our needs and the general's promise. I sent the package via FedEx so I could know when it had arrived.

The very next day, the World Trade Center towers were attacked. The FedEx tracer indicated that the general, or someone at Fort Detrick, received the package about an hour before the first attack. I remember standing next to microbiologist Christina Kellogg watching the office television as the second plane struck. We looked at each other, both wondering whether there would be anthrax throughout New York. No one knew what might happen next. I went home visibly shaken.

Would the terrorist attack stimulate our research? Would it be the break we needed? Still not a word from Fort Detrick, and in fact, I never did hear from the general, who has since retired. Not long after the air attack, anthrax-laden letters killed five people while sickening many others. The anthrax attack completely disrupted the mail system for weeks around Washington, D.C. Investigations followed, but the source of those letters remained shrouded in mystery. Several years later, the supposed perpetrator, a microbiologist from Fort Detrick, died of an overdose of a prescription drug before going to trial.

Our project received a great deal of attention in spite of limited funding, but it was still clear that USGS headquarters didn't know what to do with us. The Fisher Island orphans were still a problem. Maybe the Reston management did know what to do—simply ignore us! And that they did.

I would learn sometime later that headquarters feared the new Department of Homeland Security. There was concern they might swoop in and take over the project. To my knowledge, management never did go to Homeland Security to seek funding, although it seemed such an obvious thing to do. After the attack on the Twin Towers, we assumed our work would be high priority and that Homeland Security would be contacted. If they were, we never heard about it.

One person did take note: novelist Sarah Andrews. Sarah, a former USGS employee, had gained fame writing mystery novels featuring real geologists. Some of my friends became either villains or victims in her books. The main hero of her mystery series is a woman geologist named Em Hansen. For her ninth novel, Sarah wrote about our research. We all had different names, but our identities were obvious to everyone.

The book was titled *Killer Dust*. It told the world that we were underfunded and underappreciated. Best of all, none of us were murder victims in this novel. We enjoyed the story. It received a lot of publicity, and we received many calls from fellow geologists. Characteristically, though, there wasn't one peep from management at USGS headquarters! For a subject with so many implications and high visibility, it may seem strange to the reader that the study was a hand-to-mouth operation. Yet another example of how science sometimes works. I must say, however, that I was devoted to the USGS and they showed appreciation by awarding me a Meritorious Service Award followed later by the USGS Gene Shoemaker Award for communicating science to the public.

Diving with Hawkeye

On one occasion the USGS was asked about science projects that might make an interesting story for a public television series called *Scientific American Frontiers*. The principal player was Alan Alda, famous for his role as Hawkeye in the TV series *M*A*S*H*, and we submitted the idea that African dust would be a possible subject. Chedd-Angier Productions was the producer, and this segment of the series was being filmed in the Caribbean. Perfect! Everyone knows about dust down there, and Alda had spent many vacations in the Caribbean. We were on!

When we met, I learned Alan Alda had been reading the magazine *Scientific American* since he was fourteen years old and had a longtime interest in science. His background may explain why he played such a believable surgeon in the *M*A*S*H* series. We met in St. John, Virgin Islands. There were four of us: Garriet Smith, the microbiologist who had identified the soil fungus that affects sea fans; Ginger Garrison, who had been living in the area, knew the reefs like the back of her hand, and was committed to the dust hypothesis; Don Hickey our technician member of the A-Team; and me. We all hit it off with Alda right away. He was brilliant! He needed no script. He was Hawkeye all over again!

Alan constantly cracked jokes. I learned we are exactly the same age, that he, like me, remains married to his first and only wife, and that we both have three children. His wife is a concert clarinetist, and my wife played the clarinet in our high school band. We even subscribed to the same offbeat magazine called the *Skeptical Inquirer*, a journal devoted to debunking fake science, especially paranormal experience and its practitioners. I would later publish my story about beachrock and Atlantis in that journal.

The day after the program aired on public television, I received a call from a fellow at NASA named Gene Feldman. Gene was a young fellow in charge of a satellite program called SeaWhifs. It was the same satellite that produced the images we had been using to track dust transport across the Atlantic. He was excited about our work but couldn't offer funding. However, he could offer something else. From then on, he would provide us with some outstanding Earth images. He knew exactly what we needed. Eventually, we created a complex intertwined number of associates, but still funding was hand to mouth. Nevertheless, word was getting out and numerous diverse people would contact us. They all had interesting stories to tell.

One person who contacted us was yet another interesting fellow connected with Fort Detrick. He thought we should establish monitoring stations on the Atlantic and Pacific Coasts to establish baseline information. A baseline would let us know what microbes and chemicals were normally in the dust. Knowing what should be there would alert us if something new and potentially dangerous appeared. It was a great idea, but our microbiologists were researchers, not routine-monitoring technicians. We would need a broad network of dedicated staff technicians to maintain stations and perform analyses. Such an effort might cost millions. The funds he had to offer were not nearly enough, and I doubted USGS would back us up. We had to pass up the offer.

As mentioned, we took advantage of most opportunities. For example, Ginger Garrison had served in the Peace Corps in Africa and had many friends there. One was a schoolteacher who taught the son of the American ambassador. On vacation, Ginger went to Mali to visit her teacher friend. The purpose was to make a dusty trip up the Niger River to Timbuktu—and of course take a few samples.

Mali is a very dusty place and the source of much of the dust that crosses the Atlantic. Ginger's trip was successful and had an unusual twist. Very quickly, Ginger, the teacher, and the ambassador's son set up a dust-collecting station on the roof of the American Embassy.

Our ambassador was well aware that dust storms in Bamako, Mali, cause health problems, so he was also very interested in the work. In fact, everyone who had ever lived there knew that. But the connection got even better. The ambassador's son could ship us the dust samples in the diplomatic pouch. No charge! We now had a source of samples that could be used to characterize what was in the dust before setting sail for the Americas. These samples allowed Christina Kellogg to culture and catalogue microbes directly from Mali, a major source of dust reaching our shores. This was all before the terrorist attack of 9/11. Unfortunately, the attack put a stop to sending samples via diplomatic pouch, and not long after that the ambassador moved on to another assignment.

That wasn't the last of inexpensive opportunities. After some wrangling, we arranged to get Dale Griffin on a leg of the Deep-Sea Drilling Program. On this leg, program scientists would be drilling in the middle of the Atlantic, right in the African dust pathway. The ship sits for weeks, and sometimes months, in one place while drilling, thus providing the perfect platform for dust collection. This all happened because of good contacts with the Deep-Sea Drilling Project. My USGS colleague Terry Edgar had once been director of the program. And there was former USGS chief geologist, Steve Bohlen, who had become director of the Joint Oceanographic Institute, which helps determine which projects are to be conducted on the drill ship. Steve had been interested in our dust project before he left the USGS. Bingo.

At very low cost we had Dale Griffin flying to Brazil, where he boarded the ship for two months. On board he collected dust from the drill tower right in the middle of the Atlantic, and he could culture the microbes right there on the ship in its modern microbiology laboratory. The only downside was that no major dust storms occurred during his two-month-long stint. Nevertheless, even with little dust, he identified seventeen different species of bacteria and thirty-two species of fungi. The results were quickly published.

Later, Dale conducted a similar study from a Japanese ship in the Pacific, and after that he collaborated on a paper with a professor at the University of Turkey. The university has a hundred-meter-high tower for air monitoring. At certain times, dust storms from North Africa cross the Mediterranean, affecting Spain, Italy, France, and Turkey. Nilgun Kubilay collected and determined dust quantity and chemical composition, while Dale identified the microbes. Her paper with Dale was soon published. As noted, we took advantage of all opportunities—it was a necessity. (Note: In late 2011 Dale Griffin completed another study in the middle of the Atlantic aboard the deep-sea drilling ship. This time there were abundant dust storms, and new papers were published.)

Another opportunity was presented in the Virgin Islands. This one would thoroughly convince me that African dust had severe human-health effects. Through my friend Doug Seba, a fellow of the American Academy of Environmental Medicine, I learned of a medical doctor whose wife was very susceptible to hydrocarbons in the atmosphere. Growing up in Texas, she had lived near refineries and suffered terrible headaches. It was a normal part of life for her. To ease her pain, the doctor bought property on the windward side of St. John Island, where one can generally see the British Virgin Islands to the east, and built a home especially designed for his suffering wife. It was bare concrete, no paint, and the roof had a special coating that did not contain tar. Everything was constructed without use of chemical dyes. Furniture was unpainted wood, and there were no carpets, which contain dyes and other chemicals that affect chemically sensitive people.

For the first time, his wife enjoyed living without headaches. She thought it was heaven! But that was before the first summer dust storm arrived. When the dust blows in, you cannot see the British Virgin Islands a few miles away. This woman became our canary in the mine. She would advise us when the dust was severe, and satellite images allowed us to tell her when the storms were coming. She didn't have to look out the window to know a storm was there. She felt it.

Life was good for her in St. John—that is, until the oil fires of Desert Storm began. The dust was laden with something, probably hydrocarbons. Something in the dust affected her breathing so much that she had to be flown off the island immediately! That was a real wake-up call for me, the singular event that forever convinced me that medical effects of dust were real. Remember the seawater containing beryllium-7 in our whiting study? We already knew there were high levels of that radioactive element in our dust samples, and now we also found it in the doctor's specially designed water cistern at the St. John chemical-free home. The red silt we collected from the bottom of their cistern contained high gamma-radiation values from both beryllium-7 and lead-210. The Virgin Islands example was an important event for me, but there was yet another, and that one really indicated we were doing something important.

Back when we were researching the atomic craters at Enewetak, we worked with an Air Force scientist named Bob Couch. Bob was there to keep us educated and on track without revealing any classified information. One of the things I learned was that during the Cold War, Bob had collected "fallout" (dust samples?) from Russian atomic tests. Seismometers determined the time of events, so the trick was to quickly collect and analyze samples for short-lived radionuclides. Timing was important! Knowing the precise time between the event and analysis allowed his team to learn something about the efficiency of a bomb.

Bob could never tell us how the dust was collected, but I surmised it must be via Air Force spy planes or possibly commercial airliners that routinely transited the air lanes east of Russian territory. We really hit it off, possibly because this fellow had been the first graduate student of Christopher Kendall, the same fellow who had worked on sabkhas in Abu Dhabi while I was in Qatar.

I called Bob. He had retired from the Air Force and was working with an environmental research company. I wanted to see what he could tell me about dust. Was I doing something worthwhile? Bob had maintained all his old contacts. After making a few calls, he quickly confirmed that I was on the right track, doing something very important, and should keep on it.

From him I learned that the air-monitoring team he had worked with was dismantled after the Berlin Wall came down. That network of old boys was very unhappy about losing the program. They pointed out that when Pakistan tested its first atomic weapon we had no way to evaluate its efficiency. Learning that Bob thought our work was important gave me great confidence and motivated me to press on. It also made me think that somewhere in the government some group must be keeping an eye on the atmosphere. Maybe they were monitoring us. I certainly hoped that was true.

Funding, or Lack Thereof

As noted, the dust study was a bootleg operation right from the beginning. Management didn't know how to handle it or us! On the surface it didn't appear to be a geological study. It was a project involving multiple disciplines. It was OK to study the physics of dust movement, its mineralogy, and geologic implications, but we were after something bigger. What effects was the dust having on living organisms, ecosystems, and people? Could it be used as a carrier of bioweapons?

At this point in time, our local congressman, Bill Young, learned of our study. Young was head of the Finance Committee and a very powerful congressman. He was devoted to funding projects in our district, especially if they had something to do with national defense and the University of South Florida. It greatly helped that Peter Betzer, dean of the College of Marine Science, was on good terms with Young, and also that Peter had done his PhD dissertation on Asian dust. Peter supported our project, and it wasn't long before Young provided congressional funding to keep us going. He arranged for us to receive $700,000 for our research. That seems like a lot of money, more than I had ever seen, but the USGS had recently modified its operating procedures, which had become very different from when I first joined. Now it was expected that outside grant money would also be used to pay salary. In this case, there were five of us on the dust project. After salaries and overheads were extracted—I began using the term "ripped off"—we ended up with $65,000 to be split among five people to do the actual research. That's a paltry sum for a project with such huge implications. It was disheartening when we read in various publications about other government projects receiving millions.

The congressional funding kept us going for two years, but still we received little support from headquarters. We were also in competition with another group doing dust studies. They were examining the physics of dust transport in the western United States—a politically safe study!

Mysterious Trip to Pasadena

In 2004 I received a mysterious phone call from the secretary of something called the Agouron Institute in Pasadena, California. I had no idea what it was. Despite making several calls to people on the West Coast, I learned little. What the Agouron people said they wanted was to pay my way if I could come to Pasadena so we could sit down around a table, eat lunch, and talk about dust! Just who were these people? This was all so mysterious.

After some calls and receiving legitimate e-mails, I decided to take up their offer. I was desperate to keep the dust project going and appreciated any potential help. On the appointed day, I flew to Los Angles and took the shuttle to a nice hotel in Pasadena. There was a lot of dust in the air. Californians called it smog.

The next morning, a taxi took me a short distance to the address provided. It was a new modern office building, and the elevator took me to the appointed floor, where I entered what was obviously a recently rented office. There was a main reception area, some side offices, and a very nice conference room surrounded by large windows. It was clear that the people had just moved in. There were several people there already, and a nice lady asked what kind of sandwich I would want for lunch.

In the conference room, I met six or eight people and we exchanged the usual pleasantries. It was at this point that I learned they were mostly microbiologists and professors from Stanford and Cal Tech. Together they had developed an HIV drug that they sold to the Pfizer drug company for an undisclosed amount of money. The money had been used to create a nonprofit foundation for the purpose of supporting overlooked scientific discoveries. They were especially interested in science subjects that were not receiving the attention and funding they thought were deserved. Lying on the table were several copies of the dust papers we had published. I was impressed, intimidated, and wondering, "What's really going on here?"

At the head of the table was a Dr. John Abelson. I mentioned that I was aware of a Phil Abelson, who had been the editor of *Science* magazine for many years, and he said, "Yes, Phil is my uncle." I was impressed—and even more intimidated. The other guests included a scientist from India who also worked on dust, named Y. Ramanathan and called "Ram" for short.

Ram was from the Scripps Institution of Oceanography and was working on Asian dust and its effects on climate. I had read his publications. With him was his friend Paul Crutzen. The name meant nothing to me, but I would learn later that Paul was a Nobel Prize winner in atmospheric physics. The others were microbiologists, and it became clear that the viable microbes our group had been finding in African dust were their principal interest. I also learned that they fund research but do not receive proposals. I had heard about such people, but their existence always seemed like a scientist's dream. As mentioned in my essay on Paradigm Disease, obtaining grants often requires a long history of success in a narrow field along with long forms that require describing everything you wish to achieve along with what you have done in the past. Not this group! They weren't interested in anything done before. They were out to advance new knowledge, or so they said. I was really impressed that they had been paying attention to what my little group was doing.

Using an overhead projector, I gave a presentation describing where we were in the work and what needed to be done. We had pleasant discussions and ate our sandwiches. After me, Ram gave his presentation.

He was seeking drones that could fly thousands of miles out over the

Pacific to monitor and collect Asian dust, especially dust emanating from India that contains many man-made pollutants and affects rainfall and climate. Afterward, we had more speculative discussions. After sizing up the situation, I made it clear that they shouldn't give money to government agencies but rather to independent researchers. The original unique research they were seeking is usually not possible within government agencies, which are too programmed and overmanaged. There would be various and sundry overheads—I was reminded of how little we actually ended up with for dust research from the $700,000 congressional money. Worst of all, there would be increasing oversight. There would be restrictions on where you would be allowed to purchase supplies, which airlines to fly, and committees composed of people who would not appreciate the work. Management would have to be briefed on a regular basis.

I don't know if Ram got his drones, but he certainly has produced many scientific papers on dust since that meeting. His name pops up in news editorials on a regular basis. I remember he was highly criticized by native countrymen when he pointed out that "cooking on wood fires was a major source of pollution emanating from India." They didn't like that!

I returned without mentioning the trip to USGS management. The Agouron group was clearly impressed with the potential of dust as a carrier of disease microbes and stated that they would encourage research on the subject through other means. Whether they did I may never know. It was a unique and mysterious side adventure I will never forget. At times, I wondered if the new office was a front set up just to see what we knew or didn't know. I have tried to e-mail Ram on several occasions, as well as the secretary of the organization. There were no responses. Sometimes I feel just a little paranoid . . .

Whitings One More Time: Will It Ever End?

The twenty-first century arrived while Chuck Holmes and I were investigating the use of beryllium-7 as a possible marker for determining the age of the lime mud. Presence of beryllium-7 indicated rapid precipitation of lime mud within the water column. A paper about our work still hadn't been written or published, and whitings were still controversial subjects. During 2007, Christopher Kendall contacted me.

Chris had been evaluating a new hypothesis. He was proposing that cyanobacteria (green algae) that form intertidal algal mats, something we had both studied in our early Persian Gulf days, could be a source for oil. We now knew that whitings, thanks to the Robbins and Blackwelder and Kim Yates study, also contained cyanobacteria. Chris thought that both forms could be sources of petroleum. What he meant was that the cyanobacteria that lived many millions of years ago could be the sources of some of the oil we pump from the Earth today. The same kinds of bacteria and sedimentary processes active in the whitings probably existed back during Permian and Mesozoic time.

After burial and heating during burial, the cyanobacterial mats might even have been the source of Middle East oil. The chemistry of cyanobacteria is such that they are more easily converted, under pressure and heat, to oil and gas than most other organic substances—a hypothesis that had recently been published by three organic chemists. If ancient cyanobacteria were indeed the source of all that oil, then cyanobacteria in modern algal mats and whitings might tell us more about how all that Middle East oil had formed. Study of these processes might indeed have some practical significance after all. What if that knowledge could help us find more oil?

I remained hesitant, but Chris talked me into joining him in presenting a poster at the 2007 annual meeting of the American Association of Petroleum Geologists (AAPG), a meeting that was combined with the Society of Economic Paleontologists and Mineralogists. The poster, with all our illustrations of whitings and algal mats, basically proposed that Middle East oil is derived from cyanobacteria that grew hundreds of millions of years ago. To our surprise, our poster won the annual Best Poster Award! I can only hope that our hypothesis survives the test of time and future investigation. Science advances by disproving hypotheses, and someone is always looking for the fatal flaw in new ideas. The unique aspect of this hypothesis is that small lipid-rich cyanobacteria without hard shells will release their oily fluids if heated and compressed, while leaving no trace fossil or physical evidence behind. It is the same kind of algal cells that at the time of this writing are being grown experimentally as a potential source of biodiesel. Some major oil companies have invested millions in such research.

More Mud

At the meeting where we won the prize for our poster, Gregor Eberlie and his team from the Rosensteil School of Marine and Atmospheric Sciences presented an outstanding poster. Had I been judge, I would have voted for his poster. He had been diving in the Johnson *Sea-Link* submarine in the Straits of Florida along with a Harbor Branch biologist, where to their surprise, they found huge mounds of mud, some as much as one hundred meters high! It was lime mud that had been swept off the Great Bahama Bank during storms. They also found ledges composed of stiff mud. Whitings on the Great Bahama Bank had been producing and shedding far more mud during the last few thousand years than we had ever dreamed. That mud had been created by cyanobacteria, or at the very least, it contained cyanobacteria. It was the same mud that exited the bank through various channels during storms and made the lime-mud layers described earlier. That mud could also be the source of oil—to be generated millions of years in the future. Accumulating along the base of the Bahama Bank, the mud hundreds of feet thick demonstrates a likely model for past accumulations. Has it happened this way in the past? Well, that's what geology is all about and is why we say, "The present is the key to the past"! But in this case, we may never know for sure.

A thrill comes when science, driven by curiosity, eventually has a huge, unanticipated payoff. The little story told here is what made all those bootleg trips to the Bahamas, skirting official State Department courtesies, regulations, and bureaucratic BS, so rewarding. For me, this was right up there with accidentally finding submarine cementation in the Persian Gulf. I like to think that Nobel Prize winner Richard Feynman, one of my scientific idols, would have approved if he were still living.

At the 2008 meeting of the AAPG, I experienced another example of how science can pay off. A young fellow approached and reminded me that he had participated in a 1996 AAPG field trip along the Florida reef tract that Al Hine, Bob Halley, and I had organized. This young fellow with only a bachelor's degree in geology excitedly told me that using the ideas generated by our field course, he had discovered millions of barrels of oil in the Midland Basin in West Texas. He had actually found new reservoirs in one of the most explored oil basins in the United States. It feels good when you know your efforts haven't been wasted.

Dealing with True Believers and Atlantis

Over the years I dealt with some odd and most interesting individuals. My friends continually tell me that I attract such people because I am willing to listen to people's outlandish stories. These "interesting people" can often come up with new "alternative ideas" because they are not afflicted with "Paradigm Disease." Some, however, are just too far off the edge, but nevertheless I will politely listen. For example, I once had a call from a woman who told me she could hear hurricanes from five hundred miles away. "I have to turn my head just right to hear them," she said. Another told me, "The Bahama Bank whitings were created when the power crystal that powered Atlantis blew up." Strange people can lead you down different paths.

One such path is a crooked trail that continues today. It began back at Fisher Island when Cesare Emiliani called and said, "I'm sending a fellow over who wants you to look for Atlantis." His name was Peter Tompkins. Peter, a well-known writer, adventurer, and New Age thinker, turned out to be a most charismatic and charming fellow. We had refused his initial requests, worrying for our reputations, but we finally agreed to work with him on weekends—a fine time to dive in the clear waters off Bimini anyway.

On the first outing, Harold and I took cores from several "Atlantis" monoliths and showed not only that this beachrock was identical to the beachrock on the main swimming beach in Bimini but also that no artifacts from an ancient civilization were present. Beachrock is a very distinctive rock that forms quickly near midtide level under the sand in the tropics. Tidal fluctuation constantly forces calcium carbonate–rich waters through the sands where evaporation and off-gassing of carbon dioxide probably help stimulate precipitation of calcium carbonate. Within a few years, crystals of aragonite precipitate between the grains, soldering them together to form a very hard limestone. When sea level rises, as it has done during the past eighteen thousand years, any beachrock formed several thousand years ago has become submerged—as is the case with the "Atlantis" stones off North Bimini.

We considered the case closed, but Tompkins, not deterred by the evidence, hypothesized that perhaps Atlantians had only beachrock available as building material. Following a forensic-geology approach, then, we noted that if the stones had not been moved since they formed, then they should all contain beach sand stratification dipping in the same direction as when it formed, that is, toward deep water. Conversely, if the stones had been placed by humans and selected for best fit, then internal stratification probably would dip in different directions. We agreed then to another expedition to take seventeen oriented cores from stones in close proximity.

Tompkins enlisted a local PBS station to accompany us on the venture and make a documentary. They weren't the only ones joining us, however. When we arrived in Bimini with Captain Roy's loaded fifty-foot trawler we found a carnival-like atmosphere. Along with Tompkins were a French photojournalist, a *National Enquirer* writer, a woman who practices rebirth, and two sisters who would take underwater film footage. We donned our dive gear while most of the other folks did the opposite; clothes, it seems,

Dan Robbin coring natural beach rock off Bimini in the Bahamas. New Age thinkers thought these natural stones were the work of ancients.

hinder the force field true believers feel in this ancient spot. Soon we were being filmed by two naked women and observed by naked swimmers from the handful of sailboats that had also come to the site.

Once safely back home, we analyzed the cores, and indeed every section revealed the strata dipping consistently toward deep water. We could also trace distinctive layers of rounded beach pebbles from one monolith to the other. Surveying the site as a whole, we could furthermore explain the long rows of seemingly fitted stones being the result of erosion of exposed beachrock.

We filmed the lab work—sawing and X-radiographing the cores and explaining the origins—but unfortunately the 16-millimeter film was not up to PBS standards. I wrote an account of the Bimini expeditions that was published in *Skeptical Inquirer*, and I told the geological story in several venues. First was "Bimini's Atlantis Hoax," an article for *Sea Frontiers*, published by the International Oceanographic Foundation, which was happily received by the editor as a scientific countermeasure to the frequent requests about the seemingly unusual stones. Then, in 1980, Marshall McKusick and I coauthored a piece for *Nature*. Using carbon-14 data, we showed that the stones, which range from two thousand to four thousand years old, are far younger than the putative nine-thousand-year-old story. Indeed, the rock is even younger because the material we dated consisted of various bits of beach sand and conch-shell fragments cemented within the rock; the actual time of rock formation would have been some time after the conch shell and sand were deposited on the beach. We did not, at that time, have the new mass-accelerator dating methods that now allow dating of the tiny individual crystals that formed the rock. Furthermore, an earlier article in *Nature* showed that so-called columns on a site about two miles from the stones were made of portland cement. In the 1800s, cement was carried on ships in wood barrels; the barrels fell or were thrown overboard, and the wood rotted away, leaving a hard cement column.

Since our Bimini expeditions and research, I've observed similar formations—supposed "megaliths" and seeming rows of pavement-like rocks—world over, both in shallow-water sites such as off Vieques Island in Puerto Rico, around the Great Barrier Reef, or in the Dry Tortugas, as well as five thousand feet below the surface from the *Alvin* submarine. I simply thought that publishing the facts would stop the strange speculations and the many expensive expeditions subsidized by gullible donors. I was, of course, wrong, as there's no evidence that will change the mind of "true believers." Things quieted down for a while in the Atlantic tirade, but activity picked up again in the 1990s with William Donato's Atlantis Organization, and of course the Bahamian government doesn't mind promoting the tourist trade to the region.

Ironically, the *Skeptical Inquirer* story reached more people than any scientific article I ever published. Letters and e-mails poured in from people in faraway places I could hardly remember. Most were congratulatory, but some were not. I had awakened a sleeping monster.

The Cayce Foundation, a large organization named for Edgar Cayce, who popularized the Bimini/Atlantis myth in the 1930s, apparently unleashed teams of true believers to discredit my story. One created a Web blog that says, "Shinn is part of a government conspiracy to keep the public from knowing the truth about the stones." Books and articles are emanating from them routinely to disparage our scientific publications.

Then along came the Geological Society of America meeting in 2005, which organized a day of scientific presentations to honor the career of Bob Ginsburg. What could I present at this meeting? Fortunately, I remembered that Ginsburg's first scientific publication, in 1953, had been on the formation of beachrock at Loggerhead Key in the Dry Tortugas. That was the spark I needed. I would tell the Bimini Road Atlantis beachrock story.

Alternative science, like alternative medicine, has been gaining ground, and Intelligent Design was at the top of the list. If I could tell the facts about beachrock and Atlantis, and relate it to alternative science, I might be doing a valuable service for the scientific community. With that in mind, I used the Bimini beachrock adventure as a vehicle for telling the Bob Ginsburg story.

Afterward, Bob said my presentation was the hit of the session. That was fun, but the difficult part was that I then had to prepare a scholarly scientific presentation for the International Association for Sedimentology Special Publication that resulted from the session. The title was *The Mystique of Beachrock*. That wasn't easy to do, because I had always approached the subject with a certain amount of humor and sarcasm. I worked hard and waited three years for final publication, but eventually, the book was published in 2009. Do I think it will change the minds of the "true believers"? Certainly not! In fact, there are now books that say Atlantis is really in Indonesia, and there is a new cult of believers with a journal of their own called *Atlantis Rising.* They have annual meetings just like scientists and various trade organizations. Recently it has been proposed that Atlantis is now buried beneath a wetland area near the coast of Spain. And even since then, I have been asked to do an interview with BBC Radio on the Bermuda Triangle and the Bimini Road, another program on the subject for the Discovery Channel, and most recently a program for the National Geographic Television Channel. I wish the myth were over, but the subject has achieved a life of its own and will be with us forever.

Return to Doha

The year 2006 was wonderful! Pat and I celebrated our fiftieth wedding anniversary in Doha, a place that brought back many memories. How did it happen that we returned to the place that had so changed our lives?

Many years ago, I had given Mohammad, who was the five-year-old son of my trusted assistant Rashid when I worked there, some copies of the many photos I had taken of children and pearl divers. By then Mohammad was in his twenties. The photos were for a potential *National Geographic* article. In the 1980s he had some of the transparencies enlarged and printed on canvas so they could be hung in the Al Khor Sports Club. During the 1980s the Qatar government had built several multimillion-dollar sports clubs for the youth of Qatar. These clubs have large soccer fields, indoor basketball courts, exercise rooms, swimming pools, and other first-class facilities, including arts and crafts areas—they are for boys only. As an example of how elaborate the club is, the Al Khor club has a twenty-foot-high brass Arabian coffeepot in the main foyer.

Mohammad knew I had many more photos and recognized that the photos had historical value because they depict life in the mid-1960s. Mohammad, who had earned a PhD in physics in England, was now teaching physics at Qatar University, and he and his wife (also a teacher) have three children. Mohammad has a grown childhood friend from Al Khor named Ahmad Al-Misned, who is presently in charge of a very large project, a facility called the "Cultural Village." Its purpose is to capture, preserve, and depict a way of life that disappeared with the coming of oil wealth. I might add that the Cultural Village is equal in size to the New Orleans Superdome but is much more spread out. It was designed to re-create an old Arabian Gulf fishing village with shops and souks like in the days of old. It also includes a huge auditorium for conventions and a large outdoor arena overlooking Doha Bay. No money was spared in its construction—Ahmad Al-Misned has the backing of the emir's wife, who is a big supporter of education and cultural events.

Mohammad had shown Ahmad some of my photos and decided they were worth having for display in the Cultural Village. It took some negotiation, but Al-Misned brought Pat and me back to Qatar for ten days—all expenses paid! In exchange for the trip I donated 270 original 35-millimeter Kodachrome slides. The color transparencies included buildings, boats, falcons, pearl divers, and lots of children. Many of those children are now grown and running the country—and one was Ahmad Al-Misned himself. I will never forget that while we were dining together, he leaned over and said, "Mr. Shinn, as a child I used to chase your Land Rover when you drove through our village." Wow! That sure put me on my heels!

Al-Misned said they would reproduce those photos in a special gallery about Qatar's lost past. How could I have possibly known when I took the photos how valuable they might eventually become. Ten days in the Ritz Carlton! To top it all off, Pat and I celebrated our fiftieth at the top of the Ritz—a hotel that stands on what was a barren sabkha in 1967. As guests for our little party we invited a young Scottish biologist, Iain McDonald, and his wife. Iain worked for the company presently exploiting the largest gas field in the world. We envied them and told them they were having a unique, unforgettable foreign experience similar to what we had had in the mid-1960s.

What a change forty years had brought. Pat could hardly believe all she saw. We tried but could not find our old home on the edge of the desert, as it had been swallowed up by the ever-enlarging city of Doha. We were promised a return trip for the grand opening of the Cultural Village.

That trip to the Cultural Village wasn't my only return, or the last. I had been there just a few years earlier, when Iain McDonald brought me over to help with a coral reef problem. Iain was the environmental specialist for Qatar Gas, a liquid natural gas company sitting on North Field, one of the world's largest gas fields, at the north end of the Qatar Peninsula. Reefs offshore from their complex facility had died in 1998 and were being threatened by the huge gas development.

I had recorded the death of these same coral reefs many years earlier in 1964. What killed them back then was a severe schemal, the Arabic name for winter storms that blow in from the north. The schemal in 1964 brought freezing winds and chilled the Gulf water enough to decimate nearshore corals. It was so cold that icicles were seen in Qatar for the first time in one hundred years! However, in 1998 it was not cold water that killed corals but hot water during a worldwide El Niño. The corals had recovered from the cold of 1964 only to be killed again by warm water.

In addition to helping explain the reef problem, we also worked with a combined German and French crew doing documentation to establish a World Heritage site in the sand-dune area to the south, a place called Umm Said. The southern end of this sand dune and sabkha area forms the tense, often contested border with Saudi Arabia. Saudis and Qataries constantly watch each other through binoculars across a body of water now called the inland sea. I had core-drilled this large area in 1967, and the results were published as a chapter in a book titled *The Persian Gulf*, edited by Bruce Purser. My cores documented the only subsurface geological information available for the area. Revisiting the dunes and beaches where we had camped and swum in the mid-1960s was indeed nostalgic, especially when we endured a sand and dust storm that enveloped us for two days.

Although Umm Said was my last project in Qatar, I had returned there earlier with Peter Scholle, Mitch Harris, and Bob Halley in 1983 to help make a training film called *Arid Coastlines*. The film was financed by and produced for the AAPG. Its purpose was to help petroleum geologists recognize and more profitably explore similar subsurface areas formed hundreds of millions of years ago.

I made yet another visit in 1987 to document dune migration at Umm Said, while en route to a geological congress in Kuwait. Chris Kendall and several of his former Arabian graduate students he had trained at the University of South Carolina were with us—one was his former student Saif Al-Hajari. Saif had been a national champion soccer player with hero status and is highly revered throughout the Persian Gulf area. Saif, at this writing, heads up an educational foundation and is in charge of a dedicated educational compound composed of American universities and geological research laboratories.

One such facility in the compound is a multimillion-dollar geological research laboratory recently established by Royal Dutch Shell. That's quite a change! I operated a one-person laboratory out of an apartment for Royal Dutch in the mid-1960s. Now, not only Shell is located there, but

In 2005, Gene returned to a site in Qatar where the mineral dolomite forms milky precipitate in briny groundwater exposed in a shallow pit. Geochemists rejected the notion that dolomite can actually precipitate directly from water.

also ExxonMobil, and at least two universities, Texas A&M and Cornell University. By now there may be many additional American universities in the compound. Watching the changes around the Arabian Gulf has been most interesting for all of us who saw the area when it was mostly desert and dirt roads.

Retirement and a New Beginning

> A dashing young man named Shinn,
> Had quite a mischievous grin,
> He swam among corals,
> Of dubious morals,
> And he solved all his problems with gin!
>
> *A limerick by John Keith to celebrate Gene's retirement*

Wow! What a party it was! Friends and former mentors came from across the land. There were Ron Perkins, Paul Enos, Mike Lloyd, and Bob Ginsburg and of course the USGS office pals. It was a real send-off, and it wasn't just for me! Terry Edgar and I chose to retire on the same day, January 3, 2006. By sheer coincidence, I had signed the official retirement papers in my hotel room at the Ritz in Doha, Qatar, a few weeks earlier. They had been faxed to me at the hotel while Pat and I were celebrating our fiftieth wedding anniversary. The joint retirement party in St. Petersburg was wonderful—the USGS got rid of two old grouchy birds with one stone!

Terry Edgar had his own reasons for leaving at that time. My main reason for saying adieu was increasing bureaucracy and declining support for dust research. But there was a silver lining to which I eagerly looked forward.

Dean Peter Betzer and Associate Dean Al Hine offered me an office at the University of South Florida College of Marine Science—a five-minute walk from my old office. The new office was larger than the old one. It could hold far more of my junk—and I had plenty! So far the middle drawer of my desk isn't completely full, even though I seldom throw anything away. I became what the school calls a "Courtesy Professor." Why it's called that I don't know, but it was a life ring with only one disadvantage—no paycheck! The advantage? I now can say what I want and engage in activities such as consulting and politicking, which were not legal while a government employee. I remain a volunteer at the USGS, and there are no bureaucratic forms to fill out. Oh well, maybe a few.

The new office provided a base where I am exposed to a larger variety of disciplines and knowledge. Oceanographers, biologists, microbiologists, and fishery biologists surround me, and there are students. I consider them my family, and it is a place to go every day and avoid housework. And those

At Gene's retirement in 2006, Paul Enos, Gene Shinn, Ronald D. Perkins, and Robert N. Ginsburg re-create a scene from a 1964 field trip in Florida Bay.

"problems solved with gin"? I have to wait until at least 5 P.M. Pat says I spend more time at my new office than when I worked for a living. Even better, it provided me with a home base from which to write this book.

Committee Work

First, there are student committees, and of course everyone wants something! Most are under the illusion that the retiree has unlimited time. For me, there are two major diversions. One is the Department of Interior Bureau of Ocean Energy Management, Regulation and Enforcement Scientific Advisory Committee. That happened because of studies I had done for them in the past (when they were the Minerals Management Service) and because of my knowledge of the energy industry. Our meetings are all about offshore oil-and-gas drilling where I can draw on abundant past experiences. Interestingly, the questions about oil rigs and artificial reefs haven't changed in thirty years. Public and political resistance to offshore drilling is the same—the same concerns about blowouts and drilling mud. The industry is still hated by many, and public officials are still poorly informed. The main thing that has changed in the past thirty-five years is the price of gasoline.

The other, and most unusual, diversion is the AAPG Global Climate Change Committee. I have been an AAPG member for over thirty-two years, and for a ten-year period of time I conducted their modern-carbonate field trips and served on their distinguished lecturer tour. In spite of that long commitment, nothing prepared me for the climate change committee. We began as an ad hoc committee to prepare a formal policy statement, but the subject became so heated among the organization's thirty thousand members that it was decided that a permanent committee was needed. Our charge was to monitor the science of climate change and to advise the membership. The petroleum industry was continually under fire, because use of petroleum as fuel releases carbon dioxide—the greenhouse gas thought by many to cause global warming.

The AAPG management wisely decided to draft members with a spectrum of viewpoints—a spectrum so wide that at times I called our group a marriage made in hell. The only thing we all agree on is that we can disagree. Even though there are two sides, we were pretty much evenly split down the middle. One of the prominent members was Ray Thomasson. Remember him from way back in Midland, Texas? Ray taught me geology out in the desert of West Texas as well as the difference between a one-man yucca and a two-man yucca. I helped him collect some from the desert for his garden. Both are thorny—just like the issues we debated in our committee. The difference? It takes two men to put a two-man yucca in the back of a station wagon, while it takes a whole committee to discuss climate change!

Pete Rose, the fellow who had set us up at Fisher Island, also served as AAPG president. Pete and Ray both had something to do with this committee assignment. Chip Groat also twisted my arm to be on the committee. Chip had left the directorship of USGS and joined his old school, the University of Texas. He serves as head of the Division of Environmental Geology for AAPG.

I must say the climate committee provided a constant inducement to learn more about climate change. First, all geologists know there has been change throughout geologic time. Then there are the scary predictions about the future made by climatologists who, for the most part, know little about geology. I found it exceedingly difficult to separate science from the politics and economics. Remember, I have made mention of political effects on many former research subjects. I still remember the comment from Bob Ginsburg's wife, Helen, "It is always difficult to separate good motives from bad motives." How true! I found that to be true over and over as I traveled along the crooked road of life. Climate change is a heady subject and a good example of science and politics that occasionally went berserk.

Climate change, specifically global warming, had been on the back burner for some time. First there was the so-called ozone hole over Antarctica made famous in a book by former vice-president Al Gore. That led to the outlawing of freon used in refrigeration and air conditioners. Next the Intergovernmental Panel on Climate Change (IPCC) was created by the United Nations and made famous by Al Gore with his film *An Inconvenient Truth.* His activist approach energized many concerned citizens, leading to his being awarded a Nobel Prize along with members of the IPCC. I might add that many of the original members dropped out because a small, select group of the IPCC actually prepared reports that many of the early members disagreed with. Politics played a large role both in the way IPCC members were selected and in the way reports were written. At the center of it all was the concept of cap and trade, a way to put a tax on carbon dioxide emissions. Is it necessary, or is it just a way for governments to collect taxes? Being a committee of the AAPG, it was clear that the petroleum industry would question these motives. What was noticeable in our committee was that academics and younger members bought into the IPCC concept and stayed at one end of the table while older more conservative members fought it from the other end. I tried to sit in the middle and learn. It was interesting that we prepared a well-written document that ironically never used the term "carbon dioxide." Before the committee was dissolved by the AAPG board of governors, one of our members, Eric Barron, left to become the head of the National Center for Atmospheric Research in Boulder, Colorado. He remains thoroughly committed to climate change research, although he is now president of Florida State University. Anthropogenic-caused climate-change science has become increasingly controversial and politically motivated.

During this retirement period I have been reunited with Leighton Steward. Leighton wrote and published a climate-related book for the lay reader titled *Fire, Ice and Paradise*. It's a book about how climate has changed throughout geologic history—when there were no automobiles or coal-fired power plants. He also created a Web site titled *Plants Need CO_2*. Because of climate issues, I also was reunited with my Australian friend Bob Foster, who is very much involved with climate issues "down under." The issue has brought out the best and the worst in us all it seems.

There are new subjects, and every day there are new problems, presented in my role as Courtesy Professor. I remain hopeful that this phase of my career is not my last. There are many interesting things happening.

The Big Award

In June 2009, I received the Twenhofel Medal, the highest award given annually by the Society for Sedimentary Geology. The medal announcement was published in the bureau-wide USGS newsletter *Sound Waves* and also in the *Congressional Record* in Washington, D.C.

Gene Shinn Wins Prestigious SEPM Twenhofel Medal
By Barbara Lidz

Eugene A. Shinn, formerly of Shell Oil, then a carbonate geologist with the USGS for 31 years, received the 2009 William H. Twenhofel Medal from the Society for Sedimentary Geology (SEPM). The highest recognition given by the SEPM, the Twenhofel Medal is awarded annually to a person for his or her outstanding contributions to sedimentary geology. Gene received an honorary Ph.D. from USF in 1998 and was a commencement speaker. Since retiring from the USGS St. Petersburg Field Center in 2006, Gene has been seated as Courtesy Professor in Residence at the College of Marine Science.

Nominees for the Twenhofel Medal are chosen for having made outstanding contributions to paleontology (http://en.wikipedia.org/wiki/Paleontology), sedimentology (http://en.wikipedia.org/wiki/Sedimentology), stratigraphy (http://en.wikipedia.org/wiki/Stratigraphy), and/or allied scientific disciplines. The contributions normally entail extensive personal research, but may involve some combination of research, teaching, administration, or other activities that have notably advanced scientific knowledge in sedimentary geology. Gene has devoted his career to each of these ambitions and more and has excelled in all. As a researcher dedicated to working in the field, he is recognized as a pioneer in carbonate sedimentology, tidal flats, diagenesis, coral reef ecosystems, and in recent years, effects of transatlantic African dust on corals and human health. Gene has an innate ability often to perceive truths before others do and encourages discussion and innovative thinking. He is not afraid to speak his mind or to get on the hot seat amidst controversy. He also knows when to avoid controversy. Shinn has led numerous modern carbonate field trips to the Florida Keys and Bahamas for the AAPG, SEPM, GSA, many universities, and local societies. He has published over 150 scientific papers, produced training films, won several "best paper awards," and has received the USGS Meritorious Service Award as well as the USGS Gene Shoemaker Award for Excellence in Communications. Gene joins the ranks of other very distinguished geologists who have shaped major concepts in understanding Earth processes and history in the carbonate realm. The honor is long overdue. Gene received the award at the Society's annual meeting in Denver in June 2009. Congratulations, Gene, for a meritorious job well done!

William H. Twenhofel (1875–1957), Yale Ph.D. (1912), is regarded as the patriarch of sedimentary geology. Twenhofel was a member of the National Research Council and retired in 1945 from an illustrious academic career at the University of Wisconsin Madison. The Department of Geology at that university has been one of the top programs in the U.S. for decades. Twenhofel co-founded the *Journal of Sedimentary Petrology*, now the *Journal of Sedimentary Research*, one of the premier journals in the field of sedimentary geology.

Epilogue

Changes in Science

Much has changed since this bootstrap geologist/scientist began his ever-winding and widening road trip. During my long career, I experienced the introduction of vacuum tube room-sized computers, fax machines, desktop computers, more powerful slim-line screens, and finally the ones that fit in your hand, and all the while the World Wide Web was growing. For me the toughest change was the advent of digital cameras and the demise of Kodachrome film. Cell phones, called mobile phones in much of the world, certainly changed life as I knew it. They now link the planet with more information than any single person can ever digest. As a musician and music lover, I went from 78 rpm phonograph records to 45 rpm and then 33 rpm, and there were wire recorders leading to reel-to-reel tape, 8-track and 8-millimeter tape cassettes, followed by CDs, which at this writing are almost obsolete. Solid-state chips the size of your fingernails are here already. Of course, television and commercial jets also became accessible during my lifetime, as well as remote sensing from satellites and orbiting space stations. With all these changes, one has to ask, are we better off? And what have these changes done for science?

Clearly, these inventions have sped our ability to process data and instantly reach faraway people, in fact, faster than most of us humans can think. Airlines and businesses, as well as governments, could hardly operate without them—or could they?

As I write this, I am reminded that somehow we were able to make complicated air and hotel reservations in exotic places while traveling the world by air. That was all before computers! On the other hand, we now write scientific papers and communicate with coauthors, collaborators, and journals at the speed of light. Nevertheless, I still worry about quality. Are we making new discoveries at the rate we did before these warp-speed time-saving devices began to rule our lives? A recent book titled *The End of Science* concluded, NO! It said we are making fewer fundamental discoveries. Its thesis is that we just process more information faster and faster, and in physics, we simply add more decimal points to existing knowledge while we polish up old hypotheses and theories.

The End of Science also makes the point that fine-tuning existing knowledge requires ever more sophisticated and expensive tools. The public usually pays for these expensive tools, and the public increasingly wonders what it's gaining from it all. Interest in science is waning, and the public who pays for it with their taxes is increasingly restless. To the public, myself included, we often seem to be—as the old adage says—reinventing the wheel. I realize that this makes me sound like an old geezer who can't keep up. That may be true! I have seen much in my field of study that fits the title of that controversial book. Much of what was published before PDFs hardly exists in current literature because an increasingly younger generation of scientists now thrives on Google and other Web-based resources rather than libraries. We may be experiencing what my professor friend Pam Muller calls "Google Science Syndrome." This new syndrome likely will make Paradigm Disease more pronounced.

I remember the day when a young Bob Halley said, "I must be getting old." He was in his thirties. "Most of the papers I read on carbonate research now are not new!" My response was, "Welcome to the club!" It happens to all of us. Just try and find a researcher over forty who hasn't said that. I think I noticed it more than most because, in my case, Shell University had been far ahead of academia. That was when our new work was also proprietary, so academia didn't see the results until they were released. During those golden years of research, it took a while for such knowledge to reach the published literature. Now information literally moves in all directions with the speed of electricity.

Historians often lament, "We are doomed to repeat the past." I must also admit I have often rediscovered old information, thinking "I was the first."

As they say, "There is nothing new under the sun." Nevertheless, much has changed, and it has a lot to do with the availability of money and energy as well as with social and political changes that modify our thinking and actions.

We must now be politically correct. Imagine using dynamite to sample rocks or attacking whitings with fish poison today! I predict that the rock hammers with which we attack and sample rock outcrops will soon be legislated out of existence. Consider the fuss over stem cells or the use of animals in research. Even dissecting frogs in biology class is frowned upon, and young developing minds are encouraged to stop collecting insects. Pinning an insect collection on a board is frowned upon and forbidden in some schools.

A few days ago I asked a professor how the live corals he was keeping in an aquarium were doing. He said, "Fine." "Are you doing any research on them?" I asked. He said, "I was going to have a student do temperature studies but she refused!" Why? "She did not want to kill or stress any of them!" White rats, monkeys, and other warm furry creatures I can understand, but it's sometimes difficult to get emotional over coral polyps. But as time moves on, I'm sure there will be many more people with strong emotional feelings for polyps. The earth itself may become sacred and we may drift back toward pantheism. Things change. I guess it's called sociological evolution. I certainly have observed an abundance of it.

Moving up the Ladder, Less Fun at the Top!

Some of the changes we perceive are directly influenced by where we are in life's journey. Where are you on the research ladder? When young and beginning a career in research, our main focus is on the research. If we do good work, we are rewarded with promotions and new titles. However, as we climb the rungs we are forced more and more toward the social and political implications of our research. Inevitably, we become more concerned with how to support our work and with those starting the journey on the ladder beneath us. When you're at the bottom of the ladder, someone older and several rungs higher is fighting for the funds to keep his or her empire thriving and, thus, your project running smoothly. Inevitably, however, you reach that rung where you must find the funds to keep a project going and to support some new researchers just starting life's journey. It is a transition we all experience. During that transition, your focus and values continue to evolve. At the same time, the world around you is changing, so you must remain flexible while keeping focused on what you wish to accomplish. Unfortunately, what is happening with increasing rapidity is increasing paperwork—most generated on a computer, of course—and evolving regulations that for the most part have little to do with the research. Ask anyone in research or anyone in academia and you will hear the same story.

It is oh-so-easy to become lazy. All those mechanical and megabit electronic advances make life so much easier. Could it be that those devices actually retard the rate of discovery? Have they tended to make science less serious and more of a game? Has the game of science become more important than actual discovery of new knowledge about the Earth? Some papers I read seem more focused on the various statistical tests the researcher conducted than on the meaning and importance of the research. Why? The computer has made it so simple to do the various standard deviations and confidence-interval tests. ANOVA, anyone? Do we still think deeply about the focus of a study? Are we really looking for answers? What are we doing? Is it really becoming a game of proving with extreme statistical accuracy what simple observations should reveal? Are we just gathering data so someone else can plug it into a computer model? And then do we use the results of a model as real data to be fed into other models? Again, I sound like an old geezer. Of course, our perceptions also change with age and experience.

A reviewer recently returned a paper I had submitted for publication. He or she needed more information. The information requested? What were the dimensions of a rock hammer in several of my photographs? The hammer was there for scale, but the reviewer apparently had no feeling for the size of a rock hammer and assumed that no one else would! To satisfy the reviewer, I had to find a standard rock hammer and measure the length of the hammerhead! I was both saddened and at the same time amused by the request. All field geologists use rock hammers for scale. In the 1960s and 1970s geologists instinctively knew the size of a rock hammer, or the other favorite, a Pentax camera-lens cover. Sometimes we placed a familiar coin or a set of keys on the rocks in the photographs to provide a sense of scale. We never had to specify the exact size of these items. This experience with the reviewer sent a clear message to me that times were changing and the reviewer had never experienced geology in the real outdoor world. I wonder if the question would have been asked if I had used an iPod or cell phone for scale?

I subscribe to a government-sponsored blog dedicated to coral reef science. When the blog began about fifteen years before I began this book, it was intended to allow rapid communication between coral reef scientists. For a while it served that purpose. In time, however, conservation-minded scientists and organizations slowly began to take over. Instead of scientific questions, they called for more regulations to protect coral reefs. The question is, protect them from what? We all knew that coral reefs were on a downhill slide, but no one could pinpoint the cause or causes. Nevertheless, there were strong feelings that whatever the cause, it must be man-made. Few could accept natural change. The regulatory mind-set began to take over. We just have to regulate something! Today when I open the coral blog, called the "coral-list" which NOAA administers, I see advertisement after advertisement for jobs to manage coral reefs and other resources. That is fine, but my concern is what is it we are managing and what are we protecting the reefs from, other than the obvious physical abuses. Although these physical impacts are common and clearly damaging, they are minor compared to the overall large-scale demise of reefs that no one really understands. My point is that we should continue to do aggressive research to determine the actual causes of coral demise rather than regulate lives of citizens without scientific proof. Protective actions should be based on science rather than emotion or faith. We should approach the central questions rather than pick around at the edges. One observation serves to make my point.

As noted earlier, Florida Keys reefs began their downturn during the period of increased African dust transport. It was also about the same time that aerial and truck spraying of pesticides to control mosquitoes began to increase. Did anyone not associated with the pesticide industry do simple controlled experiments called bioassays? The assays are not difficult. One simply puts live corals in test tanks with seawater and adds known amounts of the mosquito-killing pesticide to the tanks and compares the results with control corals in plain seawater. Has anyone done such experiments, or would any agency fund such a study? To my knowledge such experiments have not been done and probably never will.

What has caused this creeping change from basic research toward management, or what I like to call "social engineering"? The last few dozen years have seen the exponential rise of non-governmental organizations (NGOs), or nonprofits. What created the need and desire for NGOs? Most likely it has been a general distrust of government and a desire by many to get back to nature and save animals facing extinction from human activities. At the same time, the organizations benefit financially. The Environmental Protection Agency was created in the early 1970s to protect the environment. It was given power to regulate and at the same time to conduct research. Combining the two functions in a single agency is bad policy and leads to conflicts of interest. Meanwhile, many citizens continue to believe that government hasn't been doing enough to save the planet.

Through various legal loopholes, one can donate to a nonprofit NGO and claim a tax deduction. NGOs are a new growth industry. There are now thousands and thousands of them, all dedicated to doing good works, whether it be to protect an individual species or animal or plant or entire ecosystems. Most NGOs do admirable work, especially those concerned with human welfare. However, I contend that some use nature, or an organism or group of organisms, as a tool (or weapon?) to appeal to political whims or prevent the activity of disliked activities or industries. Many vegetarians, for example, would like to ban the eating of meat and would if they could. I suspect there are several NGOs dedicated to that cause and some dedicated to increased meat eating. There seem to be two sides, sometimes many, to every argument or human activity.

The Environmental Protection Act has clearly on occasion been used as a lever or cause to prevent certain activities in the name of saving a species. Saving a species is highly desirable, but sometimes it appears that NGO activities are aimed simply at preventing an unpopular activity—for example, mineral extraction, coal mining, manufacturing, or oil drilling. Remember, donations to the cause are tax exempt. Yes, it is so difficult to separate good motives from bad. The reader can think of many examples.

NGOs are generally skilled at promoting their causes, and there are other NGOs that teach NGOs how to be even more skillful. They do so by pressuring politicians and help sell newspapers by providing heartrending stories that promote their cause. All of these activities, in time, result in influencing politicians, whose main goal is to be reelected. Nothing new here.

The NGO and public calls for additional management and control seldom go unheeded. Over time, jobs have been created to save this or that issue with strong public support. All of this is well and good, but what I think has suffered in the process is basic science. Will we be like Galileo or Copernicus and threatened with death? Let's hope not. I have preached too much and nothing said here will change the evolutionary course of society. We just do it to ourselves—it's in our genes. There are also nonemotional technical advances that have played a role in influencing the parade of science.

IT, Anyone?

In business and government, new computer tools have made it simple to set up programs for tracking expenditures and the progress of multiple researchers. Programs and people hired to run and program the machines to do the tracking of course siphon off funds that might have been used for actual research. One can say the same about incessant meetings. We call it accountability. For example, there was a time when around 50 percent of USGS personnel consisted of researchers. Today I am told it is around 10 percent, and this is likely the case for many other research organizations, especially universities. Bean counters (accountants) and information technology (IT) now dominate.

For example, in recent years USGS accountants instituted a program called BASIS. I never knew what it stood for and never really cared. BASIS, we were told, was meant for proposing new research and to monitor research progress. Actually, it was more about keeping track of expenditures than actual research. Working in BASIS eventually drove me to write a humorous memo reproduced here, which I titled "Back to the Future." As expected, management was not appreciative. Nevertheless, the humor and irony made its point.

Back to the Future

In the April 8, 2002, *Chemistry and Engineering News* (vol. 80, no. 4, page 42), there is a story titled, "Politics, Culture, and Science: The Golden Age Revisited," by Allen J. Bard. The story is his acceptance speech for receiving the Priestley Medal for chemistry. As the title suggests, he devotes a lot of the article to how-it-used-to-be, when kids could have Gilbert Chemistry Sets and other toys now banned for being considered unsafe. Further in his acceptance speech he says, and I quote, "The situation is approaching that envisioned by Leo Szilard in 1948 in his amusing story titled 'The Mark Gable Foundation.' In the story, the hero, sometime in the future, is asked by a wealthy entrepreneur, who believes that science has progressed too quickly, what he could do to retard this progress. The hero answers: 'You could set up a foundation, with an annual endowment of thirty million dollars. Researchers in need of funds could apply for grants, if they could make a convincing case. Have ten committees, each composed of twelve scientists, appointed to pass on these applications. Take the most active scientists out of the laboratory and make them members of these committees. First of all, the best scientists would be removed from their laboratories and kept busy on committees passing on applications for funds. Secondly, the scientific worker in need of funds would concentrate on problems that were considered promising and were pretty certain to lead to publishable results. By going after the obvious, pretty soon science would dry out. Science would become something like a parlor game. There would be fashions. Those who followed the fashions would get grants. Those who wouldn't would not.'"

That was 1948! If only Szilard could have really predicted the future. He did not envision lengthy conference calls and how e-mail and the web would keep scientists out of the lab while recycling old information. The hero in Szilard's story could have proposed that scientists submit proposals in a special computer program. It could be called BASIS. The program would require that scientists leave the lab and take training in BASIS. To further insure they spend less time in the lab, the program could be changed each year. New iterations could even be made incompatible with last year's version, thus requiring even more time in training learning the new version. To further slow down research, he could institute some kind of "knowledge bank." Scientists could then spend valuable lab time regurgitating and reformatting old information into yet another data format. Of course this effort would require the recruitment of a large information-technology force that would siphon funds away from actual research. Information technology could even become the main function of the organization. If those tactics didn't stop science in its tracks, "mandatory" training courses could be created where entire days could be devoted to non-science training courses on subjects like, how to use a credit card, internal security, health in the workplace, sexual harassment, ethics, discrimination, etc. Time allotted to different tasks could be monitored down to the quarter hour on time-and-attendance web sites and forms for sick and annual leave could be created. Who would argue that it is not in everyone's best interest to be ethical, secure, healthy, politically correct, and avoid waste? The above duties could of course be augmented to the point that scientists become so frustrated and disillusioned that they spend what little time is left writing stories like this. Others could just surf the web all day.

Although the story was intended to be humorous, it accurately depicts what many scientists experience—almost every day! It was pretty much the same point I made in my essay "Paradigm Disease." These kinds of improvements in accountability have greatly increased since I retired and left the organization. Now I see it every day at the university level. From what I hear from others, it seems to be increasing in all organizations and is continually fertilized by increasing federal and state laws and regulations.[1]

Where Is Your Permit?

Another change that is retarding natural-science investigations is the proliferation of protected parks and sanctuaries. I see no problem with preserving the diversity and beauty of natural areas. The new problem, however, is the time and effort entailed to obtain the permits required to conduct research in these domains. Remember all those natural-resource management jobs mentioned earlier? These people need something to do and regulate. Once started, groups must be created to separate regulated activities between people trained in different activities. And so on and so on.

When I began my science career, research permits were unheard of. Consider that I was even using dynamite as a research tool! I am not advocating a return to those days. The irony is that those studies and techniques were conducted in what is now a government-owned and -managed

National Marine Sanctuary! No permit was necessary back then because an organization to protect the reefs did not yet exist. Florida Bay, part of Everglades National Park, also required no permit for research, or if it did, I wasn't aware of any. Besides, we were simply taking sediment cores that do no harm. I first encountered the need for research permits when I left Shell in 1974 and set up the USGS Field Station on Fisher Island.

Everglades National Park had begun the requirement, but because we were a sister agency under the Interior Department, there was little problem. We were automatically issued a simple half-page permit good for one year. It helped that the park superintendent was an ex-USGS geologist. We used the park area in Florida Bay for training geologists about how lime sediment is formed and deposited. When the geologist in charge was transferred, there was a change—the new superintendent was a biologist from Alaska. We took the new park superintendent and Gary Hendrix, the head of their research group, on one of our field-oriented training trips so they could see exactly what we do. They were in favor of this educational use of "park resources." Still there was no problem and they liked what they saw.

But superintendents come and go on a regular basis, and when Gary Hendrix moved up the ladder to regional headquarters in Atlanta there was a noticeable change in attitude. Superintendents increasingly came from a law enforcement background, and many rangers were being sent to enforcement schools. Soon we had to prepare applications months in advance. Annual permits were no longer issued to us. Permits became needed for each individual trip, although we weren't doing anything different from before. The expanding organization felt it had to be more accountable. As the number of researchers and enforcement officers increased, the park hired a special person to evaluate research-permit applications. That person usually had little experience in research or the area, so he or she needed to circulate applications to various experts, all of which took increasingly more time.

Then, just when the person in charge began to know who was who and what each person did, he or she would be transferred and replaced by yet another person unfamiliar with the area. The learning process would start all over again, and the process became progressively more complicated with each change. It wasn't long before the superintendent was playing a larger role in the permitting process. Permits were held back if there was a means to do the research or conduct the field trips outside the park. The educational use of park resources was slipping away.

Another barrier was concerns about geologists and/or biologists walking on marine grasses or possibly taking sediment samples. It began to look like the Park Service didn't really like researchers on their marine turf.

When I last applied for a Park Service permit, it was a seventeen-page affair. The application took a full day to complete and of course was an electronic submission. After application preparation and submittal, permits often did not arrive until the day before the proposed research began. Such delays make it extremely difficult to plan fieldwork, especially when weather conditions are a concern. On one occasion, the permit was brought out by boat after we had actually begun the work, and not by a ranger but by an embarrassed researcher within the system. Why? It was explained that there was little communication between enforcement rangers and their own researchers who write the permits. Later, while conducting that same research project, I received a call from a researcher at Florida International University. He wanted me to collect some sediment samples to analyze for pesticides. Why me? It turned out that he had been unable to get a permit for his project, which was in fact funded by the Park Service.

With this background, it is easy to understand why some researchers have simply given up and said, "It's not worth the hassle, time, and stress. I'll do my research elsewhere." Unfortunately, that can mean staying in the office working with the computer. I wonder if the Park Service is happy about that—or even cares!

To further complicate matters, there are two separate National Parks in South Florida within a few miles of each other, Biscayne National Park and Everglades National Park. They require separate permits, and the enforcement rangers—they carry guns—can be difficult. Once a class of biologists was evicted from Everglades National Park at gunpoint. The class leader had forgotten to bring along the permit he had received to record birdcalls! One has to wonder if recording birdcalls does environmental harm. I know another researcher who was ticketed for taking a water sample.

A tipping-point factor hastened many of these attitude changes, and it helps explain the ever-changing attitudes. Drugs! Parts of Florida Bay were being used for airdrops to go-fast boats, the so-called cigarette boats, waiting below. Marijuana bales would be dropped from airplanes in remote parts of the bay and quickly whisked away by these fast boats. At the same time, marijuana was also being grown on remote tree hammocks deep within the Everglades. "Pot Hammock" is the official name of a tree hammock shown on government maps. The effect of this activity was that, beginning in 1977, rangers, who in the past never carried guns, were sent to law-enforcement school. When they returned, they wore pistols and/or carried M-16s! Attitudes understandably changed because of that. A few even went to the dark side and became involved with the smugglers. I know of a few who were eventually weeded out of the service.

So it is somewhat understandable that the enforcement arm of the Park Service even began pushing its own researchers aside. Whether this took place in other states, I don't know. I did, however, hear stories from geologists outside Florida. Geologists within the National Park Service are few and far between in spite of all those mountainous geologic vistas. The geology is simply considered a backdrop for the biota.

Permitting on the coral reef side of the Keys was at first even more difficult. For several years, a research permit from NOAA had to be first published in the *Federal Register*. That alone required thirty days. With paper handling at either end, the process often took about ninety days. To complicate matters, it was state of Florida agents that were enforcing permits granted by a federal agency. I rightly earned a reputation for being a thorn in their side. Finally, permitting became so difficult and convoluted that I published a little guide titled *Obtaining Research Permits in South Florida*. I did it as a service to the many university groups that wished to bring students and classes to the Everglades and Florida Keys for environmental, geological, and biological training. They had no idea whom to contact, so my little publication became very popular, or unpopular depending on which side you were on. The rules and guidelines explained in the publication unwittingly exposed a bureaucracy more complex than the agencies wished to be known. On the other hand, I made friends with those wanting permits.

Fortunately, many of these barriers to research changed when the underwater habitat *Aquarius* was installed in 1992. We all knew that the permitting process would have to be streamlined—and it was. How did we know? The National Underwater Research Program (a sister of the NOAA National Marine Sanctuary program, both under the Department of Commerce) managed the habitat and its research program. Streamlining changes indeed did take place. From then on, a permit could sometimes be obtained in less than two weeks. The permit application is now a six-page form.

Nevertheless, paperwork and/or online applications remain an unspoken deterrent to research in state and national parks. How much so is difficult to determine, but nevertheless the deterrent is there. And don't forget the State, Fish and Wildlife Service, the Corps of Engineers, and others. They all have their own requirements.

I realize this all sounds like a lot of whining, but I'm not alone and have had plenty of time to watch the bureaucracy take its quiet toll on scientific research. I know all the people involved are well meaning, but once a government system gets rolling there's no turning back—it's called Parkinson's Law. As I frequently say, "We do it to ourselves." Or, as the cartoon character Pogo put it, "I have seen the enemy and they is US!"

So where are we? Well, it's not going to get any better, but there is a way many have gotten around it all. The answer for them has been remote sensing.

Satellites brought about a major change in the way we all do our research. They are marvelous devices that have opened up entirely new research disciplines. The upside is, for the most part they require no permits unless your work has security implications. The downside is, they still don't actually get researchers in the field or allow collection of specimens for verification. Increasingly numerous researchers, especially the mathematically and computer inclined, now do their work in the comfort of air-conditioned laboratories and offices. Why go out and get dirty? Our African dust and whiting research were greatly aided by space-based images of dust clouds and milky patches of water. Satellites are here to stay, and we can't do without them. They have forever changed the way we look at Earth and how we conduct research. I hope it's for the better! Fortunately, no one or agency has yet required a permit to collect air samples.

The Adventure Never Ends

How does one end a life story without dying? Will I feel like my career is over when the story is completed? These and others were some of the thoughts and questions that plagued me when I began documenting my life journey. I could only hope that new adventures lay ahead, and indeed they already do.

In 1960 when I was measuring coral growth rates I began taking underwater photographs of several sites. I continued to take the same photos every year except when I was overseas. I continue this activity and now have fifty-two years of serial photos of the same corals, or what is left of them. My plan is to make it a sixty-year record.

One sudden and striking event in particular did offer a suitable ending. The reader may remember the adventures at Bug Spring and the oil-spill recovery device engineered by Jerry Milgram at MIT that I described earlier.

On the dark night of April 20, 2010 (ironically, Earth Day), an enormous, fiery, and explosive blowout occurred at what was called the Macondo well located in five thousand feet of water roughly fifty miles off the Mississippi Delta. The semisubmersible rig *Deepwater Horizon*, drilling an exploratory well for BP Oil Company, had made what was apparently a major new oil-and-gas discovery thirteen thousand feet below the seafloor. The approaching disaster began when drillers were in the process of removing drill mud from the well bore. The used drill mud was being pumped into the *Damon B. Bankston*, a 250-foot-long supply boat known in the industry as a "mud boat." The *Bankston* was tethered alongside the towering semisubmersible *Deepwater Horizon*. Drill mud weighs about sixteen pounds per gallon and provides the weight needed to hold back the pressure of oil and gas in the well bore. As the mud was being replaced with lighter water (about eight pounds per gallon), natural gas under pressure shot up the drill pipe and riser and onto the drill-rig floor. The cement, previously pumped into the well to plug the bore, apparently hadn't hardened sufficiently or was improperly placed. That issue will be discussed for many years. The cement failed catastrophically; ironically, management teams from BP were visiting and celebrating the successful discovery of the oil-and-gas field and the outstanding safety record of the *Deepwater Horizon*.

Unnoticed by most everyone on board, one of six diesel engines on the drill rig that operated water pumps to keep the platform stable began sucking in the highly flammable gas. The gas, serving as additional fuel, caused the engine to rev wildly. The runaway engine exploded as pressure in the well suddenly spiked, and the volatile methane ignited a series of explosions, instantly killing eleven drillers near the drill floor. Heavy steel doors

Underwater photo of a thriving brain coral near Carysfort reef in 1960. Spikes (indicated by arrows) were implanted to measure how fast coral grew around them.

Gene photographing the same (now mostly dead) brain coral at Carysfort reef in 2011. The view is approximately the same as that in the 1960 photo. One spike remains at center left.

were blown off their hinges as the rig floor became enveloped in flames. Pandemonium ensued, and apparently the automatic switch that activates the four-hundred-ton blowout preventer (BOP) at the wellhead a mile below the surface failed to function. The uncontrolled upward flow of oil and gas created a roaring inferno that soared 250 feet above the *Deepwater Horizon* while confused and panic-stricken rig workers squeezed into and hastily lowered specially designed lifeboats. Others simply leapt from the rig into the black ocean seventy-five feet below. The mud boat released its tethers and rescued 115 rig workers from floating lifeboats and those swimming to escape the inferno. In spite of valiant efforts of firefighting tugs that arrived hours later, parts of the rig melted and simply folded over. More than twenty-four hours later, the leaking and still burning rig collapsed and sank to the bottom, producing a spaghetti-like tangle of five thousand feet of riser and drill pipe, all of which were still attached to the BOP. Seawater extinguished the raging fire as the rig settled below the waves, allowing fresh crude and gas to flow freely into the Gulf from the broken riser at the wellhead. Initial reports estimated the oil gushing from the twisted and broken pipe at 1,000 barrels per day, a volume soon upped to 5,000. That volume would later be upgraded to between 15,000 and 19,000 barrels— even as high as 50,000. Every attempt to activate the BOP with remotely operated vehicles (ROVs) failed. The spewing leak would soon be the biggest spill since the Ixtoc No. 1 blowout off the Mexican coast in 1979. Ixtoc released an estimated 1.37 million barrels. Before it was capped, the Macondo spill would be close to 5 million barrels.

What would become a highly controversial decision was quickly instigated. Technicians using ROVs would inject a dispersant directly into the source of the upward-gushing oil plume. Various dispersants had been developed in the late 1960s and 1970s to deal with spilled oil, especially floating spills in marinas, but nothing nearly this massive or at this depth. Dispersants can be somewhat compared to ordinary dishwashing liquid. As a boater, I knew that a few squirts of dishwashing detergent on a small oil slick works like magic—the sheen disappears instantly! Like removing grease on a dinner plate, the dispersant causes oil to dissolve or convert to microscopic droplets that vanish down the drain. Ordinarily, dispersants are relatively nontoxic, but that can change when mixed with oil, especially processed oils. In this case, thousands of gallons of a brand of dispersant called Corexit, developed by Exxon Corporation in the 1970s, were sprayed on the surface waters as well as directly into the rising crude near the ruptured pipe. Corexit did its job and reduced the volume of oil reaching the surface, but there was great fear about what it might be doing to marine life down below. It was a calculated Solomon-like decision. Either let floating oil foul beaches and marshes, killing highly visible birds, shellfish, and mammals, or get the oil out of sight and hope it dissolves and is consumed by bacteria at depth. The correctness and consequences of that decision are still being debated. It wasn't long before dispersed plumes of microscopic droplets of oil were found at depth over large areas in the northeastern Gulf.

The first attempt to capture the oil consisted of a four-story-high steel box called a cofferdam or containment dome that was lowered to the bottom to the broken pipe from which oil was spewing. ROVs using underwater thrusters and TV images viewed in dimly lit rooms guided the hundred-ton box into position. The plan was to capture both gas and oil and funnel them to tankers on the surface. Meanwhile, the public expectantly watched the underwater drama playing out on TV images beamed right into the comfort of their own living rooms. Permanent underwater TV cameras were later set up on the seafloor, allowing multiple streaming images to be beamed into our homes. It was high drama and it was real!

Unfortunately, because of the cold temperature and high water pressure at that depth, the methane gas mixed with seawater in the confined chamber and froze, creating crystals of an icy substance called gas hydrate. The hydrate quickly filled and clogged the four-story-high containment device, which subsequently began floating upward toward the ships above with its cargo of flammable material. Fortunately, disaster was averted. That first attempt to capture the effluents had failed dramatically. Ironically, geologist Harry Roberts at Louisiana State University had tipped off several of us that formation of gas hydrate was likely. He had experimented with cone-shaped devices over natural seeps in the Gulf of Mexico to determine flow rate, but each time the devices clogged with buoyant hydrate and floated away. Nevertheless, we kept our fingers crossed, but to no avail. It should be noted that there are numerous natural seeps in the Gulf of Mexico, including the area surrounding the *Deepwater Horizon* disaster.

The next attempt to kill the well was called a "dynamic seal" or "top kill" in which drill mud was pumped into an upper port on the BOP. The mud contained a mix of pulverized rubber tires and golf balls. It was hoped the combination of the mud pressure and possibly clogging of the twisted pipe would hold back the escaping oil long enough for cement to be pumped into a lower port on the BOP and down into the well itself. This attempt to plug the well also failed.

About forty days after the blowout, and with great difficulty, the riser and drill pipe were cut off just above the BOP. While a riveted public watched on streaming TV, the diamond-studded cutoff blade being manipulated by the arm of a ROV became stuck and refused to turn! Next, a diamond-studded wire was tried, but it dulled and failed to cut. In desperation, a giant hydraulic shear was sent down to cut the pipes. The shear worked but left a ragged edge that prevented the attachment of a new riser pipe. Yet another new device was quickly fabricated. This one harkened back to the upside-down funnel device designed by Jerry Milgram.

The newly fabricated device was called a "top hat." Remember the "sombrero" first tried on the Ixtoc blowout? The BP top hat more closely resembled the device Milgram had designed, but of course his runaway oil catcher was never intended to be used a mile down, and there was still the hydrate problem. In our living rooms we watched the top hat as it was slowly lowered to the seafloor. ROVs guided it over the gushing oil. The top hat had side ports so oil could escape while it was being loosely clamped to the BOP. The fit wasn't tight, but it worked to some degree. To prevent buildup of gas hydrate, hot water and/or methanol were piped into the device. The plan was to slowly close the ports, and on day fifty-three the device was reported to be funneling around 15,000 barrels a day into a ship at the surface. Nevertheless, as the public could observe, tremendous amounts of oil still gushed into the water. Beaches and marshes continued to be oiled, and media images of oiled pelicans were heartbreaking.

The top hat did provide a temporary but partial solution. The permanent solution remained with two relief wells that had begun drilling shortly after the blowout began. Relief wells are highly technical and must intersect or come close to the blowing well several thousand feet below the seafloor so they can clog the immediate area of the producing formation with drill mud and finally cement. Directional drilling of this kind is incredibly tedious and must be precise. A relief well had been the final solution at Ixtoc, but it took nine months to kill what, up to then, had been the world's largest subsea blowout.

Some weeks later, a new top-hat-like sealing-cap device was devised that fit more snugly over the spewing riser pipe. Installation was tedious and required removal of large bolts that held the riser pipe on the BOP. The new cap was then bolted directly to the BOP, and valves left open to relieve pressure were slowly tightened. The device held! Finally, the oil had been contained. Cement was then pumped directly into the well bore. The "static kill," as they called it, had worked!

The final so-called "bottom kill" was facilitated by one of the two relief wells after waiting out a threatening storm. Beach and marsh cleanup continued and would do so for many months ahead.

Before the well was safely capped, I found myself surrounded by intense activity. The College of Marine Science at USF, my new home, became the eye of an intense oil storm. Across the hall from my office, Dr. Robert Weiseberg, our oceanographer and Loop Current guru, was working 24/7 to predict the movement of the surface oil. It was thought that the Loop Current, a strong current in the Gulf of Mexico that swings downward toward the Florida Keys and exits into what becomes the Gulf Stream, might capture the oil. It didn't happen. Almost simultaneously, our new research vessel *Weatherbird II* cruised toward ground zero, as did research vessels from the Woods Hole Institute of Oceanography in Massachusetts. They would be joined by researchers from the University of Georgia, where a young scientist, Dr. Semantha Joi, would make headlines by finding specks of oil in sediment cores from the seafloor, as would Drs. David Hollander and John Paul from our institution. For days, roving TV cameras and commentators doing interviews invaded our usually quiet halls. We were in the eye of a media storm. I shut my office door.

NOAA and the U.S. Coast Guard soon went into what many critics called a denial mode. They disagreed with findings of the academic scientists that were highly inflamed by media coverage. A new hero emerged from the disaster. Almost from the beginning, retired Coast Guard admiral Thad Allen was put in charge of operations. He remained in charge as a civilian, and under incredible pressure he did a magnificent job of directing operations and informing the public. Nevertheless, both the Woods Hole and USF reports were deemed exaggerated and inaccurate, but with further confirmation, their results could not be denied. An official government release stating that about 75 percent of the oil was gone due to evaporation and consumption by bacteria was highly contested. That short government report tarnished the newly appointed heads of several agencies and created an uproar among environmentalists. The U.S. president was highly criticized for not taking charge. Of course, there is little a president can do in such circumstances. Debate over the handling of the spill will probably continue for years, while various congressional committees and commissions rework the accumulated data. More restrictive regulations have already been enacted. What have we learned?

The extent of environmental destruction, as well as political and economic damage, by the *Deepwater Horizon* event was obvious from the onset, and it became clear that the aftermath, both environmental and legal, would be playing out for years to come. After the first thirty days of runaway oil, the mess created vast acres of floating oil that entered marshes and bird-nesting areas around the Mississippi Delta, and almost a third of the Gulf of Mexico was closed to fishing and harvesting of shrimp and oysters. Many beaches were closed, and the number of vacationers to beach areas in the northern and eastern Gulf plummeted. Millions in tourist revenue were lost even in areas that were never oiled. The lives of newly unemployed commercial fishermen hired to help in the cleanup may be forever altered. The public's nerves, raw with anger, spawned clamor for the government to fix things, to stop the gushing oil and to pay out millions in restitution, but cool heads knew full well that no part of our government was prepared to rectify disasters originating a mile below the ocean. Only the oil industry has the expertise and equipment to repair or avoid such disasters. That knowledge infuriated many who had long hated the oil industry. Only a few months earlier I had been participating in public debates favoring offshore drilling in the Gulf! From those debates I could see that feelings toward the industry were unchanged from the time I worked for Shell Oil some forty years earlier. I had often maintained that "Some people are just born to hate oil companies." Curiously, people in Europe haven't cultivated such hatred for their oil industry.

One result of the spill was the unprecedented degree of government enforcement. I'm afraid this renewed regulation fervor has likely driven yet another stake into the heart of scientific research. The regulation, based on laws created after the *Exxon Valdez* tanker spill in 1989, is called National Resource Damage Assessment (NRDA). The Oil Pollution Act (OPA) also came into play. In the name of this act, some researchers not associated with BP or with the government were stopped on public land. Government officials often confiscated scientific samples researchers had collected. They were informed they could only sample and photograph if they had an agreement with BP! Most likely, the officials were overreacting, because NRDA and OPA provide no such authority. I consider this incident as yet one more negative activity that can be added to the list.

There were many irregularities associated with the *Valdez* cleanup and those research operations, so I can appreciate what drove this regulation. I had also once been on the other side. While with Shell we were constantly pummeled, often unfairly, by the public and the media. Many hated us. I could see both sides of the arguments, and it is clear the *Deepwater Horizon* incident will be battled in the courts for years. Researchers and lawyers representing BP's interests will be battling academic and government researchers, and lawyers recruited by the government agencies will be battling each other in the courts. There will be "chain of custody" concerns just like in a real murder case. One unusual fact I learned was that photographs taken with a digital camera could not be downloaded except in the presence of an attorney. The *Valdez* lawsuits lasted twenty-one years! Most likely the truth in this case will remain somewhere between both sides and the country will suffer many consequences.

The nation as a whole has reached a tipping point. Let's hope it is for the better and, despite the unfortunate loss of human and wild life, this tipping point will with a bit of luck lead to a future of clean, safe, and abundant energy. I hope to be around to see how that future unfolds. I am also reminded of two quotes by that modern folk philosopher and baseball manager Yogi Berra, "The future ain't what it used to be" and "When you come to a fork in the road, take one." I feel I usually took the correct one.

Acknowledgments

One of my major reasons for writing this book was to inspire students to appreciate how much more there is to do, especially in the outdoors. However, after rereading the various chapters, I fear many readers will focus only on the negatives. I complained a lot. I suppose an only child is destined to do that. I can also say I evolved as I matured. It should be clear to the reader that I had a tremendous time during the early years and was not in the least political. It was toward the end as I went up the ladder with increasing maturity that I began to notice and appreciate the ever-increasing intertwining of politics and science.

What may not have come through in this book is how much I enjoyed it all. Every bit of it! Even all the haranguing about research permits and lack of funding. The discoveries have been most rewarding and made it all worthwhile. I have no regrets. It is important to have stable relationships, and every scientist should also have a close friend with whom to share his or her secrets and ideas. All of the good adventures I had were made possible because my lifelong partner stayed with me for the duration. She participated in, and enjoyed, all our adventures. At most every fork in the road we decided our direction jointly, and almost all were the right ones.

Everyone named in this book is a real person. I didn't mention the name of the goat herder (who was a fiction) and did not give the name of the general at Fort Detrick. Most everyone in the book had a hand in guiding me toward the correct forks in the road that shaped my career. I especially thank my wife, Patricia, who has stayed for the long haul, enduring many moves and adversities, and who still enjoys the adventures. My three sons,

Eugene Jr., Thomas, and Dennis, tagged along, uncomplaining, and grew up to provide wonderful grandchildren.

Professionally, it was Bob Ginsburg who recognized some latent talent and pushed me along. Ginsburg and Mike Lloyd (who passed away in 2011 after his second bout with cancer) both served as early mentors on the hard-knocks road. Then came Pete Rose, who set me up with the USGS on Fisher Island along with Bob Halley, Harold Hudson, and Barbara Lidz. Barbara has also been in for the long haul and has contributed to and enjoyed its many opportunities. Without her eagle eye in science and grammar, our many scientific publications—and this book—might never have surfaced. Finally, Peter Betzer, dean of the University of South Florida College of Marine Science, and Associate Dean Al Hine took me in after retirement and provided the rich environment and facilities that assisted the completion of this book. And of course there were my parents. The reader has already recognized that EK or Papa-San was my life hero along with Mama-San. Papa-San was a real bootstrap inspiration. And there is my aunt Virginia Allen, a retired air traffic controller who, when well into her eighties, provided input about my earliest days. How else would I have known I was at first called "Squirt"? Or that the plaque on Papa-San's old Ford said "Specially made for Earnest Hemingway"? I especially thank Bruce Purser, who edited many chapters and removed as many "Americanisms" as I would allow, making the book more readable for those on the other side of the Atlantic. I thank Betsy Boynton of the USGS, the graphic artist who arranged the many photographs for this book.

Finally, I wish to thank the staff at the University Press of Florida: Director Meredith Morris-Babb; my project editor Nevil Parker; and especially acquiring editor Sian Hunter (also a diver) who took a special interest and painstakingly edited the first draft.

I owe whatever success I have had to all the people mentioned in the book, along with those I unwittingly forgot to cite.

Notes

Chapter 1. Beginnings and Early Years

1. See Thomas Neil Knowles, *Category Five: The 1935 Labor Day Hurricane* (Gainesville: University Press of Florida, 2009).

2. In the 1980s, an eighty-year-old Key Wester named Bethel wrote the first accurate account of the Von Cosel affair. The book's title is *A Halloween Love Story*. He consulted with my father a few times, and EK said it was the first of all the many accounts that told the story the way it really was.

3. Ironically, more than thirty years later I would once again find myself playing in a symphony orchestra, this time for eight years. By sheer chance, a friend mentioned me to the conductor of the Tampa Bay Symphony. They needed extra percussionists, so I went to my first rehearsal. I was really nervous, and I hadn't read that kind of music, mostly counting measures, since college. This time, and with my maturity, I realized why so many disciplined musicians became successful at other pursuits. Nothing else I ever did required such concentration. When you are counting measures for twenty minutes, you must concentrate and stay focused. The triangle has to be dinged and the cymbal crashed at the right time. To me, it's left-brain music, that is, by-the-numbers music. Not at all like playing by ear as one does in a dance or jazz band. Jazz is a right-brain thing and you just do it naturally without much thought.

Chapter 2. The Shell Years

1. Later in my career, after I joined the U.S. Geological Survey, Harold Hudson and I perfected an underwater coring device that allowed us to determine that

spurs on all the Keys reefs, such as Looe Key Reef, had been built by fast-growing elkhorn coral. For mysterious reasons, elkhorn had also died before, about two to three thousand years ago. We also found abundant lime mud between the coral-framework builders. That there was fine-grained lime mud within the reefs—some of it cemented to form rock—could not be seen by external examination. All the surface sediment was very coarse grained and was made of particles the size and shape of which can be compared with corn flakes.

2. Dolomite is basically calcium carbonate (limestone) that contains equal amounts of calcium and magnesium. As one goes back in time through older limestone, the limestone gradually increases in dolomite content. No chemist knew how to make dolomite, and it was said to be impossible for it to form at the Earth's present surface temperature and pressure. That was the standard explanation for why dolomite increased at depth. The oil industry was interested because dolomite almost invariably has more porosity than ordinary limestone, and it's the pores that contain the oil and gas. Peter Weyl had pointed out that ordinary limestone converts to the mineral dolomite when limestone is bathed in magnesium-rich fluids. During the conversion, called recrystallization, the individual dolomite crystals that form are denser than the limestone crystals they replace. At least that was the standard explanation at that time. This internal shrinkage results in larger spaces between the resulting dolomite crystals. The increased pore space within the rock can become filled with more oil than when it was just limestone. Dolomite was indeed very important, and no one understood how it formed. If the process of formation were known, then it might give our company an edge in the competitive business of finding oil/gas.

3. In the postwar years, Shell was a technology- and science-based company led by technical managers whose main focus was exploration and production, the so-called upstream end of the business. Management had decided to take a chance on basic venture research in the hope it would have long-term benefits. In addition, Royal Dutch Shell, the parent company in Holland, had been shut down by the war, so they sent their research dollars to Shell Development Company. As a result, Shell Development, aka Shell University, had a policy of hiring only PhDs, usually the best graduates from places like the Universities of Chicago, Cincinnati, and Nebraska. It was a time when many young veterans home from the military were going to school on the GI Bill. At Shell U, these newly graduated researchers were given liberty to do basic research even if there was no direct relation to finding oil. Some did rock mechanics, some did isotope research, and some, like King Hubbert and Gus Archie, did fluid mechanics. (Hubbert's research created a problem for the company and the industry when

he correctly predicted that the United States would reach peak oil production around 1970.) Because of so much emphasis on basic research, it was understandable that geologists in the oil patch thought we needed to do more practical research. They were under pressure to find more oil. At that time, oil hadn't reached five dollars a barrel!

4. The birdseye structure is named after features seen in bird's-eye maple. The wood contains circular blebs that someone thought looked like the eyes of birds. The ones in limestone look similar, and that's how they got their name. In Europe the structures are called *fenestrae*—the Latin word for "windows."

Chapter 3. The Netherlands and the Persian Gulf

1. Grinding a rock with abrasive powders on a flat surface until the rock is smooth makes a petrographic thin section. The smooth side is then glued to a glass microscope slide. The rock is then ground down further until there is a paper-thin layer of rock fixed to the glass. The slide can then be placed under a microscope, and light is focused through the rock from below. Often a polarizer, similar to that used in polarized sunglasses, polarizes light so that certain minerals stand out from the others.

Chapter 4. Return to Stateside Shell

1. This sparker had been developed within Shell Development Company. We had used a larger version for our survey in the Persian Gulf. Instead of creating a signal using electronic boomers and air guns, this device builds up an electrical charge in a large capacitor. The high-voltage electrical charge is periodically discharged into the water through electrodes that are at the end of a long insulated cable. Think of an automobile spark plug or a stun gun. The discharge makes a loud snap, and best of all, the device could be used from a small boat, in this case a twenty-foot-long workboat. It needed only a small gasoline-driven generator, a control box, and a heavy box containing a bank of capacitors. There were also a receiver and a chart plotter. All data were recorded on the moving paper in the plotter, much like a fish finder. Navigation was by line of sight between fixed objects, such as navigation markers and onshore objects, as well as determination of angles with a sextant. There was no Decca system such as existed in the Persian Gulf, and Loran C and GPS navigation were still way in the future.

2. For the nongeologist, "vadose" is the term for the zone above the water table. Look down a well to the water below. Everything above the water is the vadose

zone. It is wet, but you cannot pump water from this zone. "Diagenesis" is the term for the transformation of soft sediment to hard rock or the transformation of one kind of rock to another kind of rock.

3. "Geopetal sediment" refers to sediment that filters into a void. Because of gravity, the sediment always falls to the bottom and forms a flat surface. Such sediment within the rock is often compared to a spirit level. When found in rock such as a geode, it provides an easy way to tell which way was up when the rock formed. For example, think of a cave. There is dirt on the floor of the cave and there may be stalactites hanging from the ceiling. Just imagine shrinking the cave down to the size of your fingernail. Now you have a geopetal structure.

Chapter 5. The U.S. Geological Survey

1. Stratigraphy, from the word *stratum*, is the study of the layer-cake way formations or rock strata are arranged in nature. The layers become younger as they are deposited upon the older, underlying layers. Because they become younger, they contain different fossils that have evolved through time. By identifying the fossils, the layers can be assigned to different rock ages, which have names. The layers are usually named after some locality or town where the rocks are well exposed and have been studied to identify the various fossils. With that knowledge, a micropaleontologist can assign rightful ages to rocks many miles distant. Even rocks on different sides of the Atlantic Ocean can be correlated with each other.

2. Many years later, I read a report written by a Washington, D.C., think tank refuting claims by Cuba that we had practiced bioterrorism using a virus to kill the tobacco crop. By then we had determined the virus arrived on a cloud of African dust that, along with other microbes, had affected many nearby islands.

3. That mind-set was one of the reasons many believed limestone could not be a major source of petroleum. I suspect this was also one of the reasons that whitings were considered to have little relevance. For many years it had been observed that fossils in shale are generally mashed flat. Shale is easily mashed by the weight of overburden because it is simply compressed clay minerals. It's the same stuff from which pottery and bricks are made. Clay initially has very high porosity and water content. Often more than 60 percent of clay is water. The minerals are shaped like tiny plates, and they tend to rotate to a horizontal position when squashed by overburden as the clay becomes buried deeper and deeper. The water is easily expelled, and layers form like pages in a book. The fossils trapped within are generally pressed flat.

Clay minerals in rare settings initially often have high organic content; thus, many shale formations in the ancient record are considered source rocks for petroleum. Under the right pressure and increasing temperature that come with burial, the organic matter converts to petroleum. With more pressure and heat, the oil converts to gas. The conversion of organic matter to oil creates pressure that expels the oil and gas into a more porous and permeable oil reservoir. Many oil-exploration ventures (called "plays") have been based on this simple principle.

Lime sediment, on the other hand, is not supposed to behave like clay. That has been a long-standing paradigm in petroleum geology. The particles are shaped differently, but the most obvious reason is that fossils in lime sediment are seldom flattened or crushed like those contained in shale. Shale is compressed clay. Lime sediment becomes cemented to form limestone more easily than clay and therefore more often preserves burrows without signs of mashing.

4. Stylolites are wispy, hair-like structures common to ancient limestone. If you have used a stall in a public or hotel bathroom you have seen stylolites, assuming that the partitions were made from limestone. As limestone becomes buried deeper and deeper, increasing pressure causes dissolution. So much limestone can dissolve that for every foot of rock there might have originally been as much as two feet. As the lime dissolves away, any nonsoluble material like clay and organics, including oil, becomes concentrated along the resulting seam. Hard limestone that forms early, like that forming on the Persian Gulf seafloor, is unlikely to develop stylolites. Soft mud, on the other hand, develops the most stylolites, depending on how deep the mud is buried.

5. Ooids are spherical grains of calcium carbonate that precipitate from seawater. They are found in places bathed by strong tidal currents that heap sand up to form ripples and sand waves more than a meter high. The constant rolling action in the strong reversing currents creates almost perfect spheres. Ooids have formed throughout geologic history and are forming today in many parts of the Bahamas. When they become rock, the rock is called oolite. The cities of Miami and Key West rest on Pleistocene oolite.

Chapter 6. St. Petersburg and a New Beginning

1. Identification and amount of these radiogenic elements are determined by counting their gamma in a gamma-ray counter. The device measures radiation using disintegrations, that is, output of gamma rays, during one minute of time. It is expressed as "dpm." Beryllium-7 has a half-life of fifty-three days. Thus, after fifty-three days of being formed in the upper atmosphere, its gamma-ray output drops to half its original amount.

Epilogue

1. During 2011, BASIS was replaced with BASIS PLUS, and for publications a new system, IPDS (Information Product Document System), has been instituted. It never ends.

Index

Page numbers in italics refer to illustrations

Eugene A. "Gene" Shinn is Courtsey Professor at the College of Marine Science, University of South Florida, St. Petersburg. Author of more than 120 peer-reviewed articles and numerous other publications, Shinn is also a fellow of the Geological Society of America, an honorary member of the Society for Sedimentary Geology, and recipient of the USGS Meritorious Service Award, the USGS Gene Shoemaker Award, and the Twenhofel Medal, the highest award given by the international Society for Sedimentary Geology.